32 Advances in Polymer Science

Fortschritte der Hochpolymeren-Forschung

Edited by H.-J. CANTOW, Freiburg i. Br. · G. DALL'ASTA, Colleferro
K. DUŠEK, Prague · J. D. FERRY, Madison · H. FUJITA, Osaka
M. GORDON, Colchester · W. KERN, Mainz · S. OKAMURA, Kyoto
C. G. OVERBERGER, Ann Arbor · T. SAEGUSA, Kyoto · G. V. SCHULZ, Mainz
W. P. SLICHTER, Murray Hill · J. K. STILLE, Fort Collins

W0043992

With 61 Figures

Springer-Verlag
Berlin Heidelberg GmbH 1979

Editors

Prof. Dr. Hans-Joachim Cantow, Institut für Makromolekulare Chemie der Universität, Stefan-Meier-Str. 31, 7800 Freiburg i. Br., BRD

Prof. Dr. Gino Dall'asta, SNIA VISCOSA – Centro Studi Chimico, Colleferro (Roma), Italia

Prof. Dr. Karel Dušek, Institute of Macromolecular Chemistry, Czechoslovak Academy of Sciences, 162 06 Prague 616, ČSSR

Prof. John D. Ferry, Department of Chemistry, The University of Wisconsin, Madison, Wisconsin 53706, U.S.A.

Prof. Hiroshi Fujita, Department of Polymer Science, Osaka University, Toyonaka, Osaka, Japan

Prof. Manfred Gordon, Department of Chemistry, University of Essex, Wivenhoe Park, Colchester C04 3 SQ, England

Prof. Dr. Werner Kern, Institut für Organische Chemie der Universität, 6500 Mainz, BRD

Prof. Seizo Okamura, No. 24, Minami-Goshomachi, Okazaki, Sakyo-Ku, Kyoto 606, Japan

Prof. Charles G. Overberger, Macromolecular Research Center, Institute of Science and Technology, The University of Michigan, Ann Arbor, Michigan 48 104, U.S.A.

Prof. Takeo Saegusa, Department of Synthetic Chemistry, Faculty of Engineering, Kyoto University, Kyoto, Japan

Prof. Dr. Günter Victor Schulz, Institut für Physikalische Chemie der Universität, 6500 Mainz, BRD

Dr. William P. Slichter, Chemical Physics Research Department, Bell Telephone Laboratories, Murray Hill, New Jersey 07 971, U.S.A.

Prof. John K. Stille, Department of Chemistry, Colorado State University, Fort Collins, Colorado 805 23, U.S.A.

ISBN 978-3-662-15430-4   ISBN 978-3-540-35233-4 (eBook)
DOI 10.1007/978-3-540-35233-4

Library of Congress Catalog Card Number 61-642

© by Springer-Verlag Berlin Heidelberg 1979
Originally published by Springer-Verlag Berlin Heidelberg New York in 1979
Softcover reprint of the hardcover 1st edition 1979

2152/3140 – 543210

Contents

Synthesis and Modification of Polymers Containing a System of Conjugated Double Bonds*

Sebastiano Cesca, Aldo Priola and Mario Bruzzone

Assoreni, Polymer Research Laboratories, San Donato Milanese, 2007 Italy

Two broad classes of hydrocarbon copolymers containing a system of conjugated double bonds randomly distributed in the chains, i. e.,
(i)ethylene-propylene based terpolymers and
(ii) triene-isobutene copolymers,
can be obtained by using coordinate and cationic catalysts, respectively. The diene functionality can be linear and inserted in the chains of (ii) or pendant-endocyclic, -exocyclic and -linear when it belongs to (i). The reactivity of the diene groups in typical reactions, i. e., with free radicals, chlorinating or oxidizing agents, during Diels-Alder additions, was investigated with both polymer and low molecular weight model compounds for elucidating the reaction mechanism of the reactions considered. With the results obtained one may conclude that in several cases the polymers investigated possess a reactivity which is qualitatively and quantitatively different from that shown by the corresponding polymers containing monoenic unsaturation. Furthermore, the peculiar reactivity of each dienic group permits the choice of functionality which is most suitable for the post-modification reaction which is desired.

Table of Contents

---

* A preliminary account has been presented at the International Conference on ,,Polymers as Reagents", San Donato (Milan), Italy, May 25-26, 1977

Preface

The discovery in the late fifties of the EPDM rubbers, prepared by random copolymerization of ethylene, propylene and a diene, stimulated extensive research on the preparation and copolymerization behavior of dienes. In these copolymerizations one of the diene double bonds is consumed, while the other is retained for the successive sulfur vulcanization. Conjugated dienes proved unsatisfactory due to their tendency to block-copolymerize and because the usual 1,4-addition mechanism leaves one double bond in the polymer backbone rather than, as desired, in the side chains.

A series of unconjugated dienes having two double bonds with different reactivity were therefore synthesized. The EPDM rubbers prepared with them on a commercial scale found many important applications especially where oxygen and ozone resistance were needed. However, one drawback limited their growth. Blends of conventional highly unsaturated diene based elastomers and of low unsaturated EPDM rubbers were not readily covulcanizable owing to the large difference in double bond concentration.

In order to overcome this drawback, new types of multi-unsaturated hydrocarbon monomers have been synthesized in recent years and their behavior in ethylene-propylene copolymerization has been extensively studied. Such monomers are characterized by the presence of one unconjugated double bond suitable for copolymerization and of a system of two or three conjugated double bonds, the high reactivity of which makes them competitive with conventional diene rubbers in sulfur vulcanization despite their low concentration.

These studies began some years ago and are still in rapid development. In addition to a considerable number of published papers, important unpublished work indicating new lines of research is in progress.

There is no up-to-date summary of the work in this area. To fill this gap and provide the last developments of the work, this paper will review the results which have been published and some as yet unpublished studies whose results are pertinent.

The synthesis of isobutene copolymers containing conjugated double bonds was accomplished only a few years ago by copolymerizing the isoolefin with suitable trienes. The results, largely unpublished, permit to compare the behaviour of these two classes of copolymers which display several analogies. The presence of the conjugated double bond functionalities gives rise to several possibilities of post-modification reactions which extend widely the utilization of these classes of synthetic elastomers.

I Introduction

Synthetic hydrocarbon polymers containing conjugated double bonds are rather uncommon. Only polyacetylenes[1, 2] and the polymers resulting from the 1,6-opening of conjugated trienes[3], such as 1,3,5-hexatriene, 1,3,5-heptatriene and 2,4,6-octatriene, have been described as possessing high molecular weight and polyconjugated double bond systems or regular sequences of two conjugated double bond systems along the chains. Coordination catalysts were used to control the chain propagation and partly crystalline polymers were obtained.

Different amounts of two conjugated double bonds were found by IR spectroscopy in polyallene obtained by means of Ziegler-Natta catalysts (e.g. $TiCl_4$ + $Al(i-C_4H_9)_3$)[4]. Depending on the synthesis conditions, polyallene can be crystalline and consists of blocks of 1,2-units and blocks of units containing vinyl groups and methylene groups conjugated with double bonds inserted in the polymer chain.

Linear or cyclic oligomers containing some conjugated double bonds can be obtained from allene or its superior homologs, eventually in the presence of 1,3-butadiene or acetylene (in the latter case co-oligomers result)[5]. Transition metal complexes, in particular low-valence complexes of Group VIII elements, were found to be the best catalysts. Dehydrogenation reactions, carried out on *cis*-1,4- or *trans*-1,4-polybutadiene and in the presence of chloranil, quantitatively yielded polyacetylene within 8–15 hours at 130 °C[6].

More recently, the dehydrohalogenation of chlorinated butyl rubber, i. e. iso-butene-isoprene copolymer containing 1.4–1.6 wt % of chlorine, essentially in allylic positions[7], was performed successfully[8]; thus polyisobutene chains carrying random groups of two conjugated double bonds were obtained.

Since it is often troublesome to indirectly introduce conjugated double bonds in hydrocarbon chains starting from polymers containing separated olefinic unsaturations, the possibility of building up macromolecules possessing conjugated diene systems in a single step is of interest.

We succeeded in obtaining ethylene-propylene-based terpolymers containing different kinds of conjugated diene systems as pendant groups of hydrocarbon chains, by copolymerizing the lowest olefins with different trienes (Table 1) in the presence of appropriate Ziegler-Natta catalysts. A common feature of these trienes is the presence of two conjugated double bonds in a polycyclic molecule, thus a third separated double bond (usually belonging to a norbornene ring) can react selectively during the terpolymerization process and the unaltered diene system enters the terpolymer chain[9]. Conversely, triene (IX) was found to give different repeat units, whereas isopropenylacetylene opened mainly the triple bond[35, 93].

Another family of copolymers containing two conjugated double bonds inside the chains was obtained by copolymerizing isobutene with conjugated trienes in the presence of cationic catalyst systems. The prevalent 1,6-opening of the triene systems affords the random enchainment of 1,3-diene systems along the polyisobutene chains, the configuration of the two double bonds being usually *trans-trans*[10].

Despite the low amount of conjugated unsaturations introduced in both the families of hydrocarbon polymers mentioned above (in general, less than 5 mol %), an enhanced reactivity was found in some reactions typical of single double bonds (e. g. radical attack, chlorination, sulfur and radical curing) and of conjugated diene systems (e. g. Diels-Alder reactions, nucleophilic and electrophilic attack, halogen addition, etc.). Apart from the interest in the parent polymers described in this work as special elastomers resistant to severe conditions – which is due to their low levels of unsaturation – several post-modification reactions can be realized by utilizing the diene function. New classes of macromolecules containing a discrete amount of polar groups (e. g. anhydride, ester, cyano, ketone, halogen, aromatic ring, etc.) inserted in hydrophobic chains and easier grafting processes and curing methods can be obtained under mild conditions.

We report firstly in this work, which completes and expands previous investigations[9], the principles involved in the synthesis of the classes of terpolymers and

Table 1. Trienes used in the synthesis of ethylene-propylene based terpolymers containing conjugated diene systems

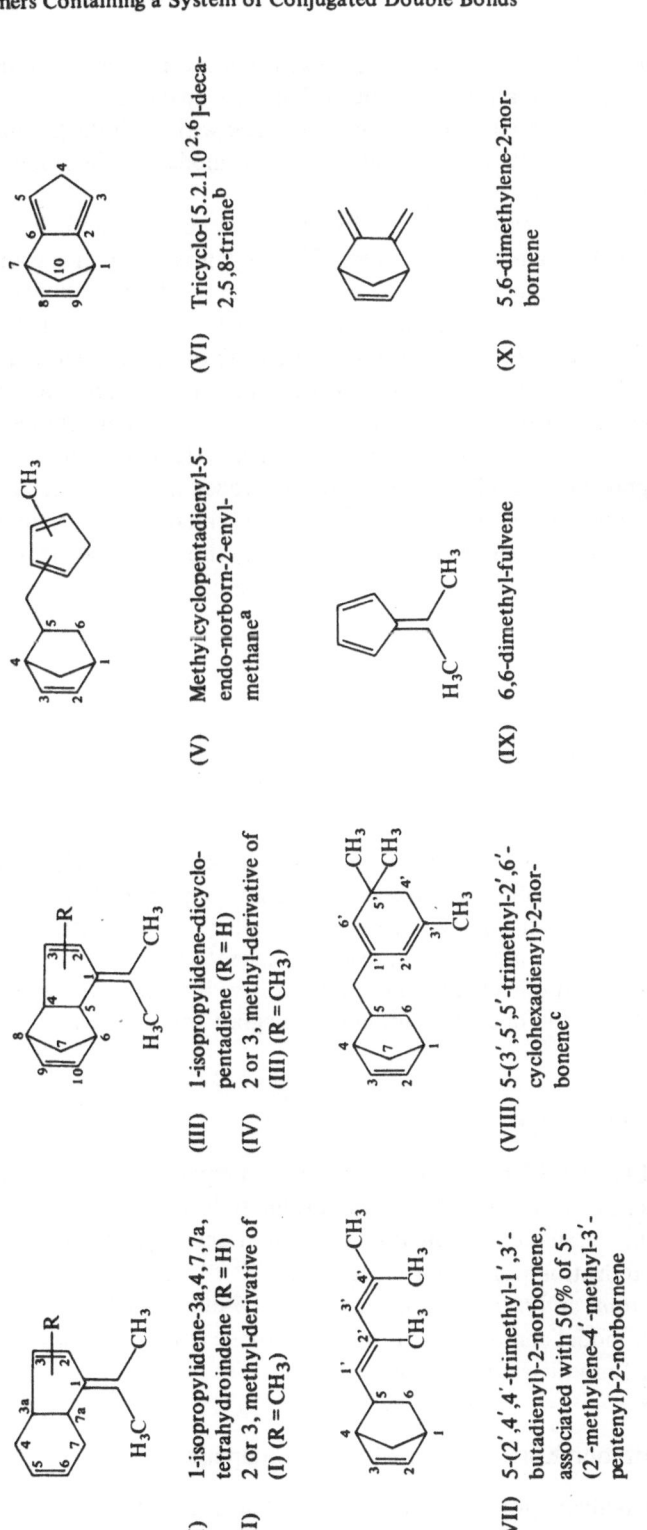

(I) 1-isopropylidene-3a,4,7,7a-tetrahydroindene (R = H)

(II) 2 or 3, methyl-derivative of (I) (R = CH₃)

(III) 1-isopropylidene-dicyclopentadiene (R = H)

(IV) 2 or 3, methyl-derivative of (III) (R = CH₃)

(V) Methylcyclopentadienyl-5-endo-norborn-2-enyl-methaneᵃ

(VI) Tricyclo-[5.2.1.0²,⁶]-deca-2,5,8-trieneᵇ

(VII) 5-(2′,4′,4′-trimethyl-1′,3′-butadienyl)-2-norbornene, associated with 50% of 5-(2′-methylene-4-methyl-3′-pentenyl)-2-norbornene

(VIII) 5-(3′,5′,5′-trimethyl-2′,6′-cyclohexadienyl)-2-norbornenec

(IX) 6,6-dimethyl-fulvene

(X) 5,6-dimethylene-2-norbornene

ᵃ Actually a mixture of positional isomers is obtained: (V,a) = 1-methyl-3-MNB-cyclopentadiene; (V,b) = 1-MNB-3-methyl-cyclopentadiene; (V,c) = 2-methyl-5-MNB-cyclopentadiene; (V,d) = 1-methyl-4-MNB-cyclopentadiene; (V,e) = 5-methyl-5-MNB-cyclopentadiene, where MNB = 5-endo-methylene-2-norbornene. Their relative abundance is (V,a) ≃ (V,b) > (V,c) > (V,d) ≃ (V,e). The last isomer is usually present as 10–15%, depending on the preparation conditions.

ᵇ Trivial name dehydro-isodicyclopentadiene.

c Present (33%) in a mixture with two positional isomers having in position 5 the following substituent:

copolymers mentioned above. Subsequently, the results concerning some post-modification reactions (i. e. radical grafting, conventional and unconventional curing, Diels-Alder, oxidation and chlorination reactions) obtained with both the groups of polymers and some low molecular weight products used as models of the reactive sites present in the chains will be presented.

New results are associated with earlier results since both contribute to an explanation of apparently different data (e. g. high efficiency in radical grafting and ability to covulcanize with highly unsaturated elastomers shown by terpolymers containing low amounts of conjugated double bonds) which come from some basic reactions involving mainly the chemistry of free radicals and of diene functionality.

The novelty of isobutene-triene copolymers does not allow us to present a complete series of results since further work is in progress[10]. However, some interesting features shown by these copolymers, e. g. radical and Diels-Alder curing, high efficiency in radical grafting, possibility of easy and controlled halogenation, allow us to expect a parallel behavior in terpolymers containing similar unsaturations and hence permit us to draw some conclusions which complete those obtained from the terpolymers study.

II Experimental Data

A) EPTMs

The preparation, physico-chemical characterization and technological evaluation of ethylene-propylene-based terpolymers containing pendant conjugated diene groups, which were named EPTMs in analogy with EPDMs according to ASTM nomenclature[11], have been described in detail in previous articles[12-20].

The experimental conditions for grafting reactions have been reported elsewhere[21] together with the methods used for analyzing graft copolymers.

B) Isobutene-Triene Copolymers

1,3,5-Hexatriene (HT) and 1,3,5-heptatriene (HPT) were prepared according to Ref.[3]. 2,4,6-Octatriene (OT) was synthesized according to Ref.[22].

Other reagents and details concerning the isobutene-triene copolymers synthesis and structural characterization were reported elsewhere[10, 23]. 2,4-Hexadiene (HD, a mixture of 30/70 *cis-trans* and *trans-trans* isomers) and 2,5-dimethyl-2,4-hexadiene (DMHD) were used as model compounds of isobutene-triene copolymers and were pure grade commercial products, distilled before use.

C) Diels-Alder Dimerization Kinetics of Some Model Compounds

Methyl-ethyl-cyclopentadiene (MEC; a mixture of positional isomers obtained by reacting equimolar amounts of sodium methylcyclopentadienide and ethyl bromide in THF: yield 60%; b. p. = 66 °C/75 mmHg; MW = 108; purity (VPC) > 99%; λ_{max} = 248 nm) and the hydrogenated derivative of $(V)^{18)}$ i. e. (V,f), were used as model compounds of triene (V) (Table 1). The thermal dimerization of MEC (solvent = n-cetane) was studied by VPC (internal standard = n-decane; C.Erba instrument, mod. GT 200; column = dimethyl ester of succinic acid (EAS), 5% on Firebrick 60 ÷ 80 mesh; length = 5 m; carrier = He, 12 cm^3/10 sec; T_{col} = 60 °C; T_{ev} = 150 °C) and by ascertaining previously, through mass spectrometric analysis, that only the dimer of MEC was formed during the thermal treatment. The experiments were carried out under argon atmosphere with solutions 0.3–1.5 M at 70 °C (±0.5 °C).

The dimerization of (V,f) was studied by UV spectroscopy (a double-beam Perkin Elmer instrument, EPS-3T mod. was used) utilizing the 254 nm absorption band (ϵ_M = 3550 l mol^{-1} cm^{-1}). n-Hexane solutions of (V,f) were used in the range of concentration and temperature, respectively, 0.97–1.22 M and 50–90 °C. Before carrying out spectroscopic measurements, the reaction solutions were diluted with known amounts of n-hexane (spectroscopic grade).

The concentrations of model compounds obtained at different times were treated according to second order kinetics and the initial rate constant was determined graphically on the conversion-time curves.

D) Kinetics of Diels-Alder Reaction Between 1,4-Naphthoquinone and Model Compounds of Some Trienes

Pure grade 1,4-naphthoquinone (NQ) was twice crystallized from acetone, and its toluene solution (0.015 M) was ascertained to be stable at 65 °C for 150 hrs through UV spectroscopy (ϵ = 46.2 l mol^{-1} cm^{-1} at 420 nm). The reaction of NQ with (V,f) was studied by measuring the absorption of the band at 420 nm, since the absorption at 254 nm, due to the dienic system of (V,f), was overlapped by the tail of the former band. The treatment of the experimental data was made according to the Wasserman approach[24, 78, 79], i. e. considering the Diels-Alder reaction as an equilibrium. Therefore the NQ concentration at infinite time was extrapolated by plotting the experimental data vs. the reciprocal of time to obtain K_e, since:

$$\frac{1}{K_e} = \frac{k_2}{k_1} = \frac{(A_0 - X_e) \cdot (B_0 - X_e)}{X_e} \tag{1}$$

where A_0 = initial concentration of (V,f);
B_0 = initial concentration of NQ;
X_e = equilibrium concentration of the adduct;
k_1 = rate constant of adduct formation, and
k_2 = rate constant of adduct dissociation.
The value of k_1 was obtained by using the relationship (2):

$$\frac{2,303}{k_1 (\alpha - \beta)} \cdot \log \frac{\beta}{\alpha} \frac{(x - \alpha)}{(x - \beta)} = t \tag{2}$$

where x is the concentration of the adduct at time t and α and β are constants involving the known values A_0, B_0 and K_e. Experimental data, plotted according to Eq. (2) and by using the least square method, yielded k_1 values, and from Eq. (1) k_2 values were obtained. An iterative procedure, carried out with the aid of a computer, showed that the formation of the dimer of (V,f) was practically negligible (less than 0.5% within 100 hrs) when the experimental concentrations of the reagents and adduct were used.

The reactions of (VI,a; i.e. the dihydro-derivative (norbornane) of triene VI) with NQ is very slow at T < 80 °C and only a few experiments were carried out with this model compound because of difficulties in realizing spectroscopic measurements at higher temperatures.

E) Reactions of Free Radical Initiators, Oxygen or Cl₂ with Some Model Compounds

n-Heptane solutions (0.25 M) of different model compounds, i. e. the dihydro-derivatives (norbornane) of triene (III) and (VII), named (III,a) and (VII,a) respectively, 2,4-hexadiene (HD) and 2,5-dimethyl-2,4-hexadiene (DMHD), were allowed to react with 50% (molar basis) of benzoyl peroxide or azo-bis-isobutyronitrile at 75 °C. The course of the reactions were followed by VPC (column = EAS, 5%, 3 m; T_c = 60 or 115 °C; T_{ev} = 180 °C; carrier = He, 10 cm³/10 sec) and UV spectroscopy (Perkin Elmer Instrument, Mod. 402). The absorption bands used were at 253 (III,a), 234 (VII,a), 228 (HD) and 242 (DMHD) nm. Preliminary tests showed that, in the absence of free radicals, model compound solutions, maintained in the dark under inert atmosphere at ·75 °C, were completely stable. The reaction mixtures, after elimination of the solvent, yielded oleous residues which contained (Mass spectrometry) mainly fragments of initiator bonded to one or two molecules of model compound.

The experimental conditions concerning the kinetics of interaction of tert-butoxy radical with some model compounds were reported elsewhere[25].

n-Heptane solutions (3 M) of (III,a), 2-ethylidene-norbornane, HD and DMHD were put in the presence of pure O_2 at 75 °C under a pressure of 800 mm Hg. The reaction course was followed by VPC and UV analysis as described above. After removal of the solvent the oleous residue was subjected to elemental analysis, $\overline{M}_n$ determination (Mechrolab instrument mod. 302), mass spectrometry (LKB instrument, mod. 9000), ¹H-NMR (Varian H-100 instrument) and IR (Perkin Elmer instrument, mod. 21) analyses. CHCl₃ solutions (0.2 M) of HD were reacted at −20 °C for ten minutes with equimolar amount of Cl₂ (dissolved in CCl₄). Chlorination products were analyzed by VPC-MS (column = EAS 5%, 3 m: T_c = 95 °C: T_{ev} = 180 °C, carrier: He) and ¹H-NMR analysis.

F) Cross-linking Density Measurements

Overall cross-linking density data (ν_t) of sulfur vulcanized (V)-EPTM and ENB-EPDM were obtained from stress-strain measurements performed under equilibrium conditions and using the Mooney-Rivlin and Guth-Einstein equations. Chemical cross-linking density data (ν_c) were obtained through the empirical relationship found by

Baldwin et al. for ENB-EPTM[26]. The quantitative evaluation of the type of cross-links was carried out according to Ref.[27].

In the case of peroxide-cured EPTMs or EPDMs, the cross-linking density (ν_t) was obtained through swelling measurements (in n-heptane at 30 °C) and using the interaction parameter proposed in Ref.[26], whereas in the case of isobutene-triene copolymers the interaction parameter suggested by Sheenan and Bisio[28] was adopted.

G) Chlorination of Isobutene-Triene Copolymers

1,3,5-Hexatriene-isobutene copolymer (HTI) was dissolved (4 wt%) in n-heptane (or $CHCl_3$) and allowed to react at 0 °C in the dark with a slight molar excess of Cl_2 (dissolved in CCl_4) with respect to the content of conjugated unsaturation. After ten minutes the reaction solution was treated with a KI solution, then was washed with H_2O and the polymer recovered by coagulation from methanol. The dried polymer was analyzed for chlorine and conjugated double bond (UV) content[10].

The HCl addition to HTI was carried out at −20 °C by bubbling at atmospheric pressure an excess of anhydrous HCl into a stirred $CHCl_3$ solution of copolymer. Samples of the polymer were withdrawn periodically from the reaction mixture and analyzed, after purification, for chlorine and conjugated unsaturation content. The chlorine content was obtained by burning the polymer according to the Shöniger method.

III Results and Discussion

A) Synthesis of Hydrocarbon Polymers Having Conjugated Unsaturation

1. Ethylene and Propylene Copolymers Contaning Different Systems of Two Conjugated Double Bonds

We previously described [9, 12–19] in detail the synthesis and several properties of ethylene-propylene-based terpolymers containing diene systems as random pending groups. For the sake of completeness we outline here some general conclusions derived from the previous work.

The high nucleophilicity of conjugated diene systems require coordination catalysts for the synthesis of the corresponding terpolymers (EPTMs)[11], which display the lower acidity compatible with their activity in order to avoid the formation of very high molecular weight macromolecules and/or cross-linked material (gel). However, it should be noted that a certain degree of Lewis acidity is always associated with Ziegler-Natta catalysts involved in the homo- or copolymerization of higher α-olefins. In fact, "neutral" systems based on organic salt of transition metals, e.g. alcoholates, acetylacetonates among others, associated to trialkylaluminum derivatives, are unable to copolymerize propylene with ethylene and eventually a third monomer[29]. This occurs because V compounds are reduced by AlR_3 to V(O) which is known, as V(II) species, to be inactive in α-olefins copolymerization and in α-olefins syndiotactic homopolymerization, whereas, the homopolymerization of ethylene is still possible with low valence V species[9]. Therefore, catalyst system based on organic salts of V and R_2AlCl achieve the best compromise in EPTM synthesis between catalytic activity and acidity of the system. But the utilization of dialkyl-Al-halides as components of coordination catalysts can, in principle, generate cationic species able to promote the polymerization of olefins and diolefins. In fact, R_2AlCl, which are very weak Lewis acids, can be quickly activated by proton donors, e. g. traces of water[30] or allylic hydrogen atoms[31], and also by other Lewis acids[32] (e. g. transition metal halides) which may be present in the polymerization system.

As a consequence of the situation depicted above, terpolymers based on ethylene and propylene are expected to contain different amounts of branching[33] or gel[34], even when they contain only single olefinic unsaturation. It has been calculated[34] that in the case of terpolymers based on 5-ethylidene-2-norbornene (ENB) the gel point is passed at about 0.2 wt % of ENB.

However, in the case of EPTM synthesis the formation of gel can be avoided by adopting low terpolymer concentration, low termonomer concentration and very efficient stopping procedures[9, 18, 19].

The structure of the triene used as termonomer plays a crucial role in reducing side reactions and obtaining high utilization of termonomer. The following features are of outstanding importance:

(i) The presence of an isolated unsaturation, very reactive toward coordination catalysts (as it is norbornene double bond) allows termonomer utilization higher than 60%. This result was actually obtained with trienes (III), (V), (VII) and (VIII) whereas conjugated trienes such as (I), (VI) and (IX)[35] exhibited utilization near to 35% for termonomer contents lower than 5 wt %.

(ii) A shielding effect on the conjugated diene system, exerted by alkyl substituents, reduces, for steric reasons, the possibility of undesired side-reactions. Furthermore, these alkyl groups introduce allylic hydrogens which are favored sites for radical hydrogen abstraction processes during curing or radical grafting[21, 25].

(iii) The structure of the conjugated double bond system, i. e. cyclic or exo as occurs, respectively, in the case of (V), (VI) and (I)–(IV), (VII), influences markedly the reactivity of the same systems during the EPTM synthesis. In fact, the presence of a cyclic diene system in trienes (V), (VI) involves two bis-allylic hydrogen atoms which are known to be very active in ionic mechanisms[19, 31], while the cyclopentadiene ring is very prone to undergo Diels-Alder reactions[36].

The latter points are very important when EPTM post-modification reactions are taken into account.

Less attention has been paid to ethylene copolymers (ETM)[14, 15] containing the same trienes used in EPTM synthesis. Their study was performed mainly with the aim of confirming the polymerization mechanism and the reactivity sites of the trienes involved in the synthesis of the corresponding EPTMs. The results confirmed the expectations, since trienes (I) and (III) randomly enter polyethylene chains through the opening of the isolated unsaturation and hence ETMs contain pending conjugated double bonds. The same catalysts used to prepare EPTMs were found suitable to synthesize ETMs free of gelled material. Even though high fractions of trienes (I) and (III) have been introduced in polyethylene chains (up to 35 wt %), the main properties of high density polyethylene, i. e. high crystallinity, insolubility in hydrocarbon solvent at $T < 120$ °C, high MW, were observed in ETM when the content of comonomer was only a few percent. Propylene-triene copolymers cannot be obtained with this type of catalyst based on organic salts of V and R_2AlCl, since the largest amounts of active centers originate from ethylene molecules[37]. The small tendency of propylene to homopolymerize under the mild conditions of terpolymerization is also typical of all the termonomers, even when they possess the strained norbornene double bond[14, 16].

2. Copolymers of Isobutene with Conjugated Trienes

The introduction of conjugated diene functions in polyisobutene chains has been recently accomplished through an indirect approach which utilizes the dehydrohalogenation of chlorinated butyl rubber, i. e. randomly chlorinated poly(isobutene-co-isoprene), by means of basic agents[8].

The high and specific reactivity of the conjugated diene system imparts to the polymer some peculiar characteristics and allows interesting possibilities of application[38].

We succeeded, few years ago[39], in obtaining the direct insertion of conjugated diene groups in polyisobutene chains by copolymerizing cationically the isoolefin with low amounts (less than 5 mol %) of linear tri-conjugated trienes such as 1,3,5-hexatriene and 2,4,6-octatriene. The advantages of the one-step approach are quite evident considering the simplification of the process which is based on the possibility of copolymerizing directly different conjugated trienes which, in turn, are available from the dehydration of sorbyl alcohol[3] and the catalytic dimerization of butadiene[22]. Furthermore, the concentration of different conjugated diene groups in the polyisobutene chains can be easily varied over relatively broad range.

Linear, triconjugated trienes homopolymerize under the action of cationic catalysts by the 1,6-opening of the trienic system (this possibility was briefly reported in Ref.[3]) and under appropriate conditions interchain reactions can be avoided, thus high molecular weight soluble polymers can be obtained.

The trienes used by us in isobutene copolymerization investigations, the structure of the repeat unit of the triene and some properties of the resulting copolymers are listed in Table 2. The first three trienes of Table 2 yield soluble copolymers, even when relatively high concentration of conjugated unsaturations enter the chains. In the case of allöocimene the 4,7-opening is prevalent and only a few conjugated groups are present in the corresponding copolymer since a pendant isobutylene group and an olefinic unsaturation (in the chain) are prevalently generated by this triene. In contrast gelled material was always obtained when 2,5-dimethyl-1,3,5-hexatriene was used.

Table 2. Cationic copolymerization of isobutene with linear conjugated trienes

Type	Comonomer	Structure	Prevalent opening of trienic system	Structure of the repeat unit	Copolymer properties[a]
1,3,5-hexatriene	$CH_2=CH-CH=CH-CH=CH_2$		1.6	$-CH_2-CH=CH-CH=CH-CH_2-$	Soluble, random, high MW
1,3,5-heptatriene	$CH_2=CH-CH=CH-CH=CHCH_3$		1.6	$-CH_2-CH=CH-CH=CH-CH-$ $\quad\quad\quad\quad\quad\quad\quad\quad\quad\;\; CH_3$	Soluble, random, high MW
2,4,6-octatriene	$CH_3CH=CH-CH=CH-CH=CHCH_3$		2.7	$-CH-CH=CH-CH=CH-CH-$ $\;\;\; CH_3 \quad\quad\quad\quad\quad\quad\;\; CH_3$	Soluble, block, high MW
2,6-dimethyl-2,4,6-octatriene (alloöcimene)	$CH_3-C=CH-CH=CH-C=CHCH_3$ $\quad\quad\;\; CH_3 \quad\quad\quad\quad\quad CH_3$		4.7	$-CH-CH=C-CH-$ $\quad\;\; CH \quad\; CH_3$ $\quad\;\; \parallel$ $\quad\; C(CH_3)_2$	Soluble, partial block, medium MW
2,5-dimethyl-1,3,5-hexatriene	$CH_2=C-CH=CHC=CH_2$ $\quad\quad CH_3 \quad\quad\;\; CH_3$		1.6	$-CH_2-C=CH-CH=C-CH_2-$ $\quad\quad\;\; CH_3 \quad\quad\quad\quad CH_3$	Prevalent gelled[b]

a   High $\overline{MW}_v$, i.e. $> 3.10^5$; medium $\overline{MW}_v$, i.e. $< 10^5$.
b   The structure of the repeat unit was determined on the soluble homopolymer.

To obtain high MW, soluble copolymers of isobutene with triconjugated trienes, one needs to use $EtAlCl_2$ or $AlCl_3$ in CH_3Cl, polymerization temperatures ranging between −50 and −80 °C, and an appropriate reaction medium able to dissolve both the monomer and the polymer. The last points is of crucial importance in obtaining soluble copolymers. In fact, by working in slurry, as in the case of butyl rubber synthesis, the resulting copolymer is highly cross-linked. Mixtures of linear, aliphatic and chlorinated hydrocarbons, e. g. n-heptane and CH_3Cl, completely suppress the formation of gel. The explanation for this may be the limited diffusion of the monomers when the copolymer separates in the reaction medium, so that the active sites also attack the unsaturation present in the polymer chains forming cross-links as a result. In a homogeneous phase the attack of the monomers is preferred and the formation of gel can be irrelevant. Similar explanations have been invoked for the synthesis of highly unsaturated butyl rubber performed under homogeneous conditions[40].

Systematic studies were carried out on copolymers of isobutene with 1,3,5-hexatriene (HTI) and 2,4,6-octatriene (OTI)[10]. The structure of the triene units present in the copolymer chains was determined by means of ^{13}C-NMR by making reference to the cationic triene homopolymers and to the same copolymers previously hydrogenated. In the case of HTI there are only traces of repeat units descending from the 1,2- and 1,4-opening of the trienic system, the 1,6-enchainment being prevalent; some cyclic structures are also present[10]. The cationic homopolymer of 2,4,6-octatriene results from the prevalent (ca. 80%) 2,7-opening of the monomer, while the remainder unit results from 2,5-addition. A similar situation also seems to be present in OTI[10].

The monomers distribution in HTI chains has been investigated by ^1H-NMR and ^{13}C-NMR on both hydrogenated and raw copolymer. Both the methods show the presence of triads attributable to III, IIE (EII) and EIE (where I = isobutene and E = 1,3,5-hexatriene), and their intensity is very close to the calculated random distribution of monomeric units[10].

The study of the monomer distribution in OTI was more difficult than in the case of HTI, even when the content of triene was relatively high and the copolymer was hydrogenated. Homopolymeric sequences of both isobutene and the triene are prevalent in OTI, while the signals due to alternating triads are weak. These results were also confirmed by cross-linking experiments reported in Sect. B.2.d. Of course, the continuous introduction of low amounts of 2,4,6-octatriene during the course of the copolymerization allows the reduction of the length of the triene blocks and the approach of a random distribution. Table 3 shows the dependence of the conjugated double bond content on the concentration of HT in the feed for typical polymerization conditions. The possibility of deducing the reactivity rations of the monomers from the classical relationship of copolymer vs. feed composition is prevented by the occurrence of some cyclization and gelling reactions when the triene content increases. However, the same information can be deduced from the analysis of the monomer sequences obtainable from ^{13}C-NMR spectra[10].

The introduction of a vinyl group in position 1 in a molecule of butadiene markedly increases its reactivity, i. e. that of 1,3,5-hexatriene, toward isobutene. In

Table 3. Copolymerization of isobutene with 1,3,6-hexatriene (HT)[a]

Run n°	EtAlCl₂ (mmole/l)	HT in the feed (mole %)	Conversion (%)	$[\eta]$[b] (dl/g)	$\overline{M}_v \cdot 10^{-5}$ [c]	HT in the copolymer[d] (mole %)
1	1.67	0.68	52	2.57	6.00	0.32
2	1.67	1.24	34	2.39	5.50	0.41
3	2.50	1.74	26	2.01	4.20	0.89
4	2.91	2.00	35	1.90	3.90	1.10
5	2.91	2.50	38	1.86	3.75	1.29
6	2.58	2.40	25	n.d.	–	1.65
7	3.33	2.60	31	2.23	4.80	1.80
8	3.33	3.00	31	n.d.	–	1.95
9	4.17	4.50	31	1.98	4.15	2.51

[a] Conditions: CH_3Cl = 40 cm³; n-heptane = 40 cm³; isobutene = 40 cm³; T = −70 °C; time = 20 minutes.
[b] In cyclohexane at 30 °C.
[c] $[\eta] = 2.65 \cdot 10^{-4} \, \overline{M}_v^{0.69}$, Ref. T. G. Fox, J. P. Flory, J. Phys. Colloid. Chem. *53*, 197 (1949)
[d] Obtained by measuring the absorption band at 235 nm of isooctane copolymer solutions and using the value of ϵ = 22.400 l/mole · cm of 2,4-hexadiene.

fact, on the basis of the reactivity ratios of butadiene and isobutene[41], 10 mol % of butadiene in the feed yield a copolymer containing less than 0.1 mol % of diene.

B) Post-Modification Reactions of Hydrocarbon Polymers Containing Conjugated Unsaturations

The modification of polymer chains with different reactions, subsequent to their synthesis, has been performed for some time because the utilization of certain macromolecules, e. g. elastomers, or the introduction of desired chemical functions, is possible only with chemical operations which are usefully realized in steps separated by the polymer synthesis. This situation is widely met in polymer grafting which has recently aroused a good deal of interest[42−44].

The availability of polymers containing conjugated diene unsaturations is of interest in this context, since the reactivity of two conjugated double bonds is generally higher and wider than that of a single double bond. Therefore, new possibilities of polymer post-modifications are expected from investigations of copolymers containing different conjugated double bond systems, such as those reported in Table 1 or those coming from the copolymerization of 1,3,5-trienes with isobutene (Table 2). Indeed, interesting results have been obtained, as will be shown in the following sections, even though some difficulties were also encountered as a consequence of the high reactivity inherent in some conjugated unsaturations. For instance, prolonged storage, e. g. one year, can produce a significant increase in Mooney viscosity and also gel formation, since self-vulcanization is a second order reaction in dienic system concentration. An effort was made during our work to define with the aid of low molecular weight products (model compounds) the origin and the extension of side-reactions (as far as these undesired processes in the synthesis step are concerned, see Sect. III A) or concurrent mechanisms existing in the main reactions investigated, i. e. radical grafting, radical and sulfur curing, Diels-Alder reactions.

1. Radical Grafting

Radical grafting of styrene and acrylonitrile copolymer (SAN) onto elastomeric backbones (e. g. polybutadiene, obtained as a latex in the presence of potassium persulphate and 2% of divinyl benzene) is of great practical importance since ABS resins are produced by this method[45]. However, even though the grafting efficiency of these processes is rather low (usually below 20%), the amount of graft polymer formed is sufficient to impart interesting properties such as impact resistance to the resulting composite.

Improved ABS-similar resins can be obtained by grafting SAN onto ethylene-propylene-diene terpolymers (EPDMs) which contain, usually, a much lower degree of unsaturation than polybutadiene, thus achieving higher thermal-oxidative resistance[46]. However, only EPDMs containing a sufficient amount (7–10 double bonds per 1,000 C atoms) of reactive unsaturations, e.g. ethylidene or isopropylidene groups, display a grafting efficiency sufficient to bring about compatibility of the glassy phase with the rubbery one and hence satisfactory final properties.

a) Grafting of EPTMs

Recently, we found[21] that EPTMs based on (I)–(VI) (Table 1) yield good grafting efficiencies even when the content of unsaturation is an order of magnitude lower than that of the excellent EPDM based on 5-ethylidene-2-norbornene (ENB) (Fig. 1).

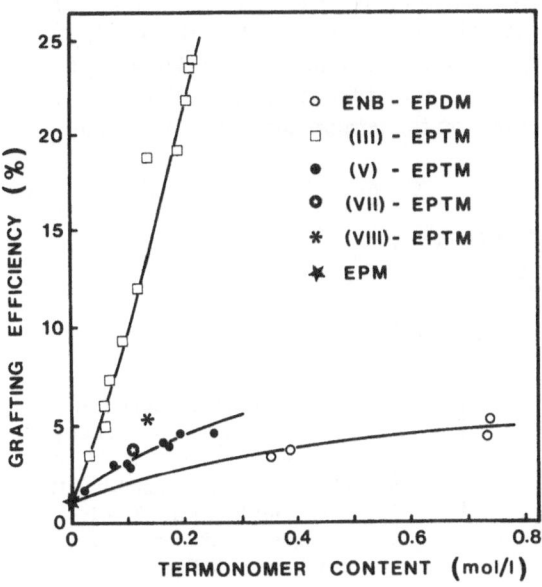

Fig. 1. Grafting efficiency vs. termonomer content for some ethylene-propylene based terpolymers. Conditions: styrene-acrylonitrile (molar ratio 1.5 : 1) = 2.6 mol/l; terpolymer = 25 ÷ 30 g/l; BPO = 8.67 mmol/l; solvent = n-heptane-benzene (1 : 1 by wt); T = 70 °C; time = 12 h

These results were achieved by working in hydrocarbon solutions with low concentration of elastomer (2.5–3.0 w/v %), an excess of styrene and acrylonitrile (molar ratio 1.5 and 2.5 mol/l) and in the presence of benzoyl peroxide (BPO) as radical initiator. The latter component is of crucial importance since other free radicals, known to act prevailingly through a double bond addition mechanism (e. g. $(CH_3)_2CCN$, originated from azo-bis-isobutyronitrile (AIBN) decomposition)[47–49], give poor grafting efficiencies (Table 4). Conversely, free radicals which prefer to abstract allylic hydrogen atoms (e.g. t-butoxy radical) give a high degree of graft. These results have been confirmed by experiments carried out with model compounds of termonomers (III), (VII) and ENB allowed to react with free radical initiators (AIBN or BPO) under conditions similar to those adopted for EPTM grafting (Sect. B.1.b)).

We believe, on the basis of this evidence, that the grafting mechanism of SAN onto (I)–(IV)-based EPTMs occurs through a prevalent hydrogen abstraction mechanism. This conclusion agrees with the presence of an elevated number of allylic hydrogen atoms on the molecules of (I)–(IV) and (VII) whose dienic system, being sterically hindered by the presence of methyl groups, can hardly add bulky free radicals. In fact, when (I) was allowed to interact with AIBN under very drastic conditions, we observed that only 33% of the triene was modified by the attack of cyanoisobutyric radicals after 24 hours at 70 °C. However, the volatile products (4.1% of the initial amount of (I)) always contained the $C(CH_3)_2CN$ group. Only a minor amount (2.1%) consisted of products involved in abstraction or donation of hydrogen atoms. Furthermore, in the presence of styrene no copolymer with (I) was obtained and the triene behaved as a retarder of styrene homopolymerization (reduction of both yield an MW)[14].

Because of the excellent behavior of (III)-EPTM (Fig. 1), we have investigated the grafting of SAN onto this elastomer with the aid of BPO, since its decomposition temperature allows sufficiently fast SAN formation at relatively low temperatures (70–90 °C).

As is evident in Fig. 2, the ethylene content and the MW of EPTM do not influence the grafting efficiency, even though the impact strength of the resulting material benefits by EPTMs high ethylene content and high MW[21]. These experimental observations agree with the assumption that the grafting reaction occurs, essentially, under the conditions adopted, on the termonomer units inserted in EPTM chains.

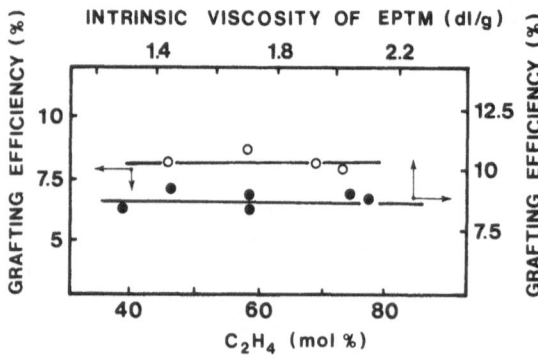

Fig. 2. Dependence of grafting efficiency on ethylene content and MW of (III)-EPTM. Conditions: T = 83 °C; BPO = 6.41 mmol/l; (III) = 0.055 mol/l; EPTM = 22 g/l; monomers = 1.85 mol/l; other conditions as in Fig. 1

Table 4. Grafting of styrene and acrylonitrile onto (III)-EPTM[a] in the presence of different free radical initiators[b]

Run n°	Initiator (type)[c]	Initiator (mmole/l)[d]	T[d] (°C)	Conv. (%)	EPTM in final polymer (%)	Grafted EPTM (%)	[η] of ungrafted EPTM[e]	Degree of graft[f] (%)	Grafting efficiency[g] (%)	[η] of SAN[h, k] (dl/g)
1	AIBN	5.7	85	83.3	13.3	45.0	1.58	18.0	2.7	0.67
2	BPO	6.8	85	84.7	13.2	74.0	1.01	74.1	11.2	0.58
3	TBPH	7.8	85	82.8	13.4	77.0	1.16	47.0	7.3	0.58
4	TBPB	6.9	120	87.1	12.8	86.8	0.80	81.0	11.9	0.37
5	DTBP	12.1	130	77.0	14.2	69.7	0.91	65.0	10.7	0.44

[a] C_2H_4 = 67 mole %; (III) = 0.056 mole/l; [η] = 2.07 dl/g.

[b] Conditions: EPTM = 20.6 g/l dissolved in n-heptane/toluene (1:1 by vol.); styrene/acrylonitrile = 2.54 mole/l, mole ratio = 1.49; time = 6 hours.

[c] AIBN = Azo-bis-isobutyronitrile; BPO = Benzoyl peroxide; TBPH = ter-butyl-5-ethyl-perhexanoate; TBPB = ter-butyl perbenzoate; DTBP = di-ter-butyl peroxide.

[d] Initial concentration and decomposition temperature chosen to obtain 5.6 mmole/l of decomposed initiator after 6 hours of reaction.

[e] In toluene at 30 °C; ungrafted EPTM was extracted with boiling n-hexane.

[f] (Weight of graft SAN). 100/(Weight of EPTM).

[g] (Weight of graft SAN). 100/(Weight of total SAN).

[h] In MEK at 30 °C; ungrafted SAN was extracted with boiling acetone.

[k] SAN formation occurring under azeotropic conditions.

The conversion of styrene and acrylonitrile to copolymer is very fast in aliphatic hydrocarbons, but relatively slow in aromatic hydrocarbons (Fig. 3) since both SAN and graft copolymer separate from the former solvents as soon as copolymerization starts. The higher the rate of copolymerization, the lower the grafting efficiency; furthermore, it decreases progressively with the reaction time (Fig. 4). However, the number of chains of SAN grafted by each EPTM chain in n-heptane is lower than

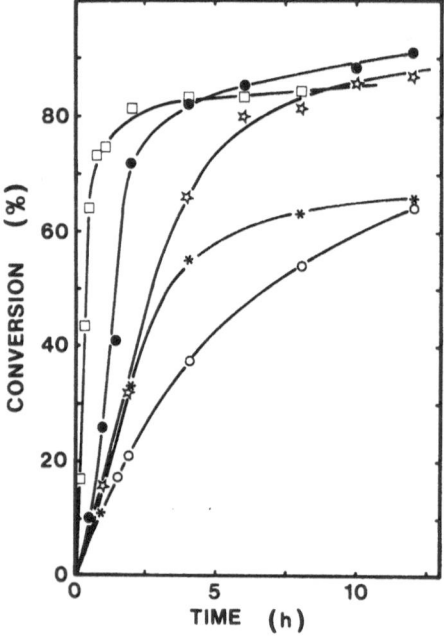

Fig. 3. Conversion of styrene and acrylonitrile as function of time in different solvents. Conditions: as in Fig. 2; (○) = benzene; (∗) = toluene; (□) = n-heptane; (☆) = n-heptane-benzene (1 : 1 by wt) mixture; (●) = n-heptane-toluene mixture

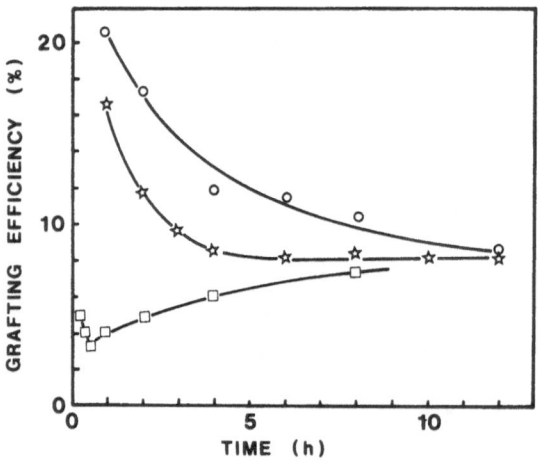

Fig. 4. Grafting efficiency of (III)-EPTM as function of time and type of solvent. Conditions and symbols as in Fig. 3

that grafted in aromatic solvent, while the weight is the same after eight hours of reaction. This means that the chains of SAN are longer when they grow in aliphatic hydrocarbons, as confirmed by direct MW evaluations on free SAN[21].

Under the experimental conditions adopted, about 5–15% of the initial EPTM remains ungrafted, but it contains C_6H_5COO fragments due to recombination of benzoyloxy radicals with the radicals formed previously on the elastomer.

Usually asymptotic diagrams of the grafting efficiency as a function of time are obtained. Meanwhile, the degree of utilization of terpolymer unsaturation is rather low (Fig. 5). This does not mean necessarily that all the potential sites of attack are consumed after prolonged reaction times. Several hypotheses can be put forward to account for this result, e. g. terpolymer cross-linking, reduction of terpolymer reactivity after separation of the graft copolymer, decrease of monomer concentration at high conversion. Further work is necessary before one can reject or confirm these hypotheses since many intercorrelated parameters (e. g. concentration of radical species, physical state of the macromolecules involved in the grafting process, monomer diffusion, preferred solvation phenomena, distribution of molecular masses) simultaneously influence the grafting process.

An attempt was made[95] to simulate the complex system of SAN grafting onto (III)-EPTM initiated by benzoyl peroxide in benzene solution. The approach was based on the kinetic analysis of a reaction scheme involving 54 reaction steps which led to 26 differential equations for the variables of interest, i.e. grafting efficiency and rates of monomers conversion.

The data obtained were compared with the experimental results obtained in a previous work[21, b] and a satisfactory agreement was found since the trends shown by Figs. 3 and 4 were reproduced.

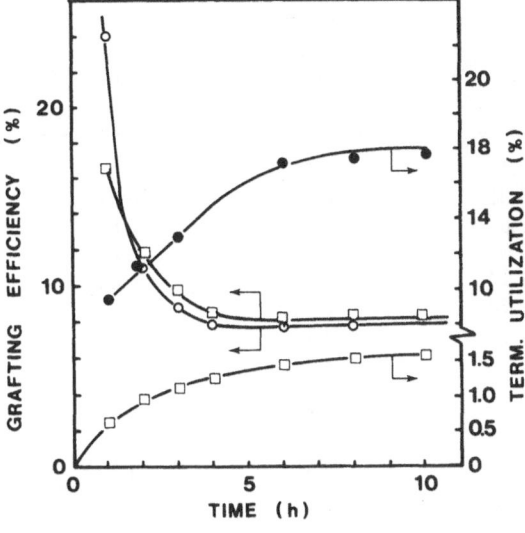

Fig. 5. Dependence of the grafting efficiency and termonomer utilization on the reaction time. Conditions: as in Fig. 3 (●) = (III)-EPTM, (III = 0.05 mol/l); (□) = ENB-EPDM (ENB = 0.61 mol/l)

The basic points of this study were: 1) the use of the Gear's stiff type method for the direct integration of the system of differential equations without the recourse to the steady state assumption [96]; 2) the experimental determination of the rate constants for the most crucial steps (i.e., benzoyloxy radical addition and hydrogen abstraction from the model compound of (III) and hydrogen abstraction from the ethylen-propylene moiety), when not derivable from the literature data.

The complexity of the reaction scheme considered stems from the variety of radicals involved in the initiation and reinitiation steps (e. g., two different radical initiators, benzoyloxy and phenyl; three different sites of attack in EPTM chains; see Sect. 3.2.a), and in termination steps which include couplings between growing chains, rubber and solvent radicals.

The agreement between calculated and experimental data was excellent at low monomers conversion while, at high conversion, an exponential decrease of the termination and propagation rate constants as function of the conversion was introduced in order to obtain the best fitting of the two series of data. This restriction of the original model and mathematically expressed by relationships of type $k_{t,p} = k_{t,p}^0 \exp(-a,b.\,C)$, where a and b are empirical parameters and C is the fraction of monomers polymerized, is consistent with the following experimental facts: a) graft-copolymer separates from the reaction mixture; b) SAN chains aggregate as separated phase in the final composite; c) potential EPTM reaction sites are, very likely, occluded by the coil of the elastomeric chains existing in the reaction solvent.

Table 5 gives a complete picture of the grafting process in terms of concentration of reactants, products and intermediate radicals at the beginning and after 10 hours of reaction.

Table 5. Model simulation of grafting of styrene-acrylonitrile onto (III)-EPTM. Calculated distribution of reactants and some radical species pertaining to various competing reactions vs. reaction time

Reactants and Intermediates (mole/l)	Reaction time (h)		
	0	2	10
BZP ($\times 10^3$)	6.4	5.0	1.9
EP	0.31	0.31	0.31
(III) ($\times 10^4$)	12.6	10.9	8.6
Solvent (SH)	9.0	9.0	9.0
Styrene (M_1)	1.14	0.88	0.44
Acrylonitrile (M_2)	0.70	0.54	0.28
BZO$^\cdot$ ($\times 10^{12}$)	0.0	4.7	2.5
Ph$^\cdot$ ($\times 10^{10}$)	0.0	8.8	8.2
EP$^\cdot$ + (III)$^\cdot$ ($\times 10^{11}$)	0.0	6.6	2.7
$M_1^\cdot$ + $M_2^\cdot$ ($\times 10^8$)	0.0	3.6	4.8
EPTM$_1^\cdot$ + EPTM$_2^\cdot$ ($\times 10^9$)	0.0	3.9	2.1
S$^\cdot$ ($\times 10^{10}$)	0.0	7.9	13.7

The study of the influence of the grafting temperature and of reagent concentrations indicates[21] that the amount of SAN grafted onto EPTM increases when mild conditions of reaction are adopted, i. e. low temperature, low concentration of initiator and monomers, but high concentration of EPTM and use of aromatic solvents.

However, since free radicals generally lack selectivity in attacking substrates or vinyl monomers, grafting efficiencies lower than 100% are to be expected.

The comparison of some properties of ABS and ATS resins, the latter being based on both ENB-EPDM and (III)-EPTM, is reported in Table 6.

b) Investigations with Model Compounds

In Sect. B.1.a) we have summarized some previous results[14] obtained from the study of the interaction of triene (I) with AIBN performed with the aim to test the reactivity of the conjugated function present in (I) toward cyanoisobutyric radicals.

More complete investigations were carried out successively with the aid of model compounds of the conjugated unsaturations present both in some EPTMs and iso-butene-triene copolymers. The norbornene double bond of trienes (III), (VII) and of ENB was selectively hydrogenated and the corresponding derivatives were allowed to react with AIBN or BPO under conditions similar (but more drastic, in order to reduce the conversion times) to those adopted for EPTM grafting. The reactions were followed by VPC and UV spectroscopic analysis to obtain, respectively, the total conversion of model compounds and the amount of addition reaction involving the conjugated diene systems. Preliminary experiments had shown that model compounds were stable for long times under the conditions adopted (T=75 °C) and in the absence of free radical initiators. Furthermore, only oleous products were recovered from the reaction mixtures and hence no chain reactions leading to solid polymers occurred in these experiments. Mass spectrometric analysis of the oleous products revealed that they are essentially formed by dimeric derivatives of model compounds containing one or two fragments of the initiator used.

The results obtained for the model compound of triene (III), named (III,a), are shown in Fig. 6. It is evident that BPO is more reactive than AIBN, and that in the first 30 minutes the H abstraction mechanism is by far predominant. However, at high conversions the addition mechanism is still important. Conversely, the conversion curves obtained with AIBN by VPC and UV analysis are very close, and this means that the addition mechanism is prevalent in this case.

When the conjugated function is linear, as in the case of (VII,a) (Fig. 7), AIBN displays a reactivity similar or higher than that of BPO, but the latter initiator acts through the allylic H abstraction, whereas the former adds the conjugated double bonds.

The model compound (2-ethylidene-norbornane, NB) of the reference dienic termonomer ENB is less reactive toward BPO than compounds (III,a) or (VII,a) because AIBN attacks NB only after prolonged reaction times (Fig. 8). This result explains the experimental observation described by Meredith[46], according to which ENB-EPDM cannot be grafted by SAN in the presence of AIBN. In contrast, Figs. 6 and 7 allow us to predict that EPTMs containing linear conjugated unsaturations can be grafted even in the presence of the less favorable AIBN.

Table 6. Some properties of ABS and ATS resins[a]

Properties (Type of elastomer)	ABS (Polybutadiene[b])		ATS (ENB – EPDM[c])		(III)-EPTM[d]		
Elastomer (%)	12	20	14	16	10	12	15
$[\eta]_{MEK\ SAN}^{30\,°C}$ (dl/g)	0.45	0.45	0.45	0.52	0.55	0.54	0.50
Degree of graft (%)[e]	ca. 55	ca. 55	100	60	65	70	70
Flow rate (g/10 min at 200 °C)	6	2	2.3	3.5	2.0	1.5	1.0
Izod impact (J/m)	70	260	450	500	150	250	360
Modulus (MPa)	2800	2200	2200	2100	2700	2400	2200
Rockwell Hardness (R)	109	103	102	101	109	106	103
HDT (°C)	89	89	90	92	94	93	90
Yield strength (MPa)	55	45	50	47	57	53	49
Tensile strength (MPa)	45	38	47	43	50	48	47
Elongation at break (%)	14	22	10	13	9	12	14

[a] Molded at 4.0 MPa.
[b] Structure: trans-1,4 = 58%; 1,2 = 22%; cis-1,4 = 20%.
[c] Containing 0.58 mol/l of ENB.
[d] Containing 0.050 mol/l of triene (III).
[e] Defined as in Table 4.

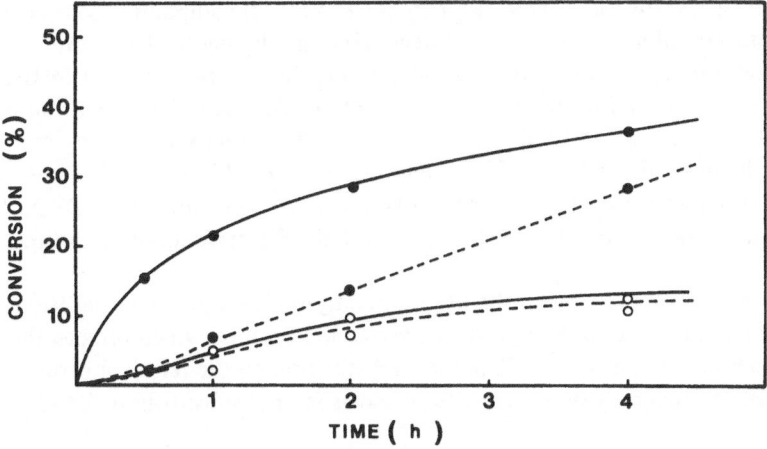

Fig. 6. Reactivity of (III,a) toward two different free radical initiators as function of time. Conditions: solvent = n-heptane; T = 75 °C; [(III,a)] = 0.25 mol/l; (III,a)/free radical initiator = 2.0 (molar ratio); (●) = BPO; (○) = AIBN; (———) = VPC data; (– – – –) = UV spectroscopy data

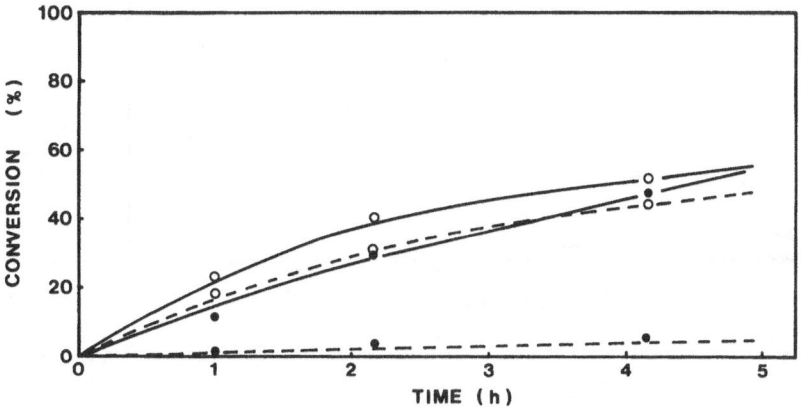

Fig. 7. Reactivity of (VII, a) toward BPO and AIBN as function of time. Conditions and symbols as in Fig. 6

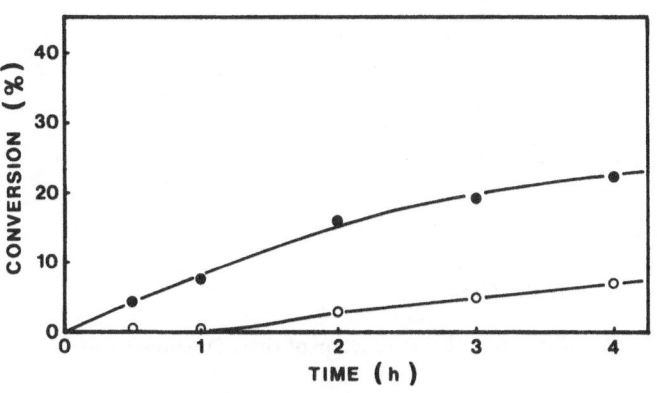

Fig. 8. Reactivity of NB toward BPO and AIBN as function of time. Conditions and symbols as in Fig. 6

These kinds of investigations were also extended to model compounds of trienes used in isobutene copolymers, i. e. 2,4-hexadiene (HD, simulating the dienic unsaturation introduced by 1,3,5-hexatriene) and 2,5-dimethyl-2,4-hexadiene (DMHD, simulating the unsaturation introduced by 2,5-dimethyl-1,3,5-hexatriene). The corresponding results are reported in Figs. 9 and 10 and they are rather similar to those obtained with the linear diene (VII,a) (Fig. 7) discussed above. The more linear the diene function, the higher the reactively of AIBN in comparison with that of BPO, and a significant fraction of AIBN acts through an allylic H abstraction mechanism (Fig. 9).

Quantitative data are available in the literature only for methyl radical addition to conjugated dienes relative to the hydrogen abstraction reaction. According to the results obtained by Szwarc et al.[50–52], methyl substitution on the terminal carbon atom of the diene system decreases reactivity, whereas methyl substitution elsewhere

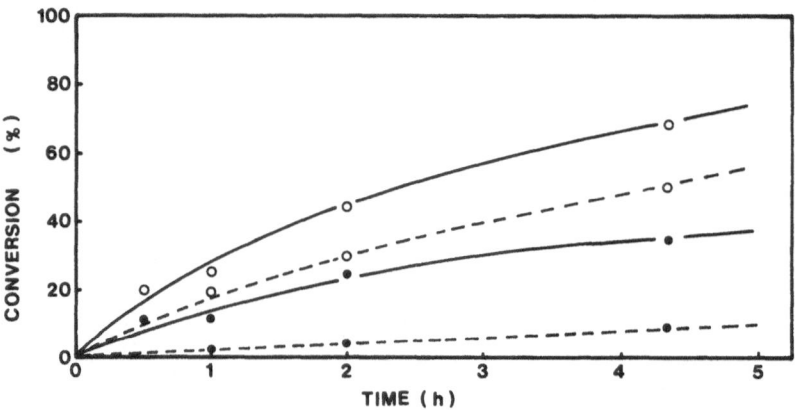

Fig. 9. Reactivity of HD toward BPO and AIBN as function of time. Conditions and symbols as in Fig. 6

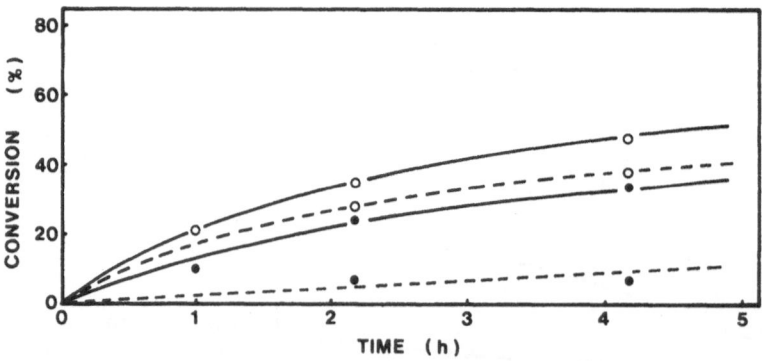

Fig. 10. Reactivity of DMHD toward BPO and AIBN as function of time. Conditions and symbols as in Fig. 6

increases it. The calculation of radical-localization energies of some conjugated dienes[53] suggests that the reaction center is on the terminal carbon atom, and this conclusion may explain Szwarc's results and ours, at least in the case of linear conjugated diene functions. Steric hindrance seems to play a less important role in this kind of reaction.

Indirect, quantitative information on the behavior of ethylene-propylene based terpolymers in the presence of t-butoxy radicals have been obtained through kinetic studies carried out on model compounds of triene (III), (V), (VI) and on the reference diene ENB. The model compounds denoted (III,a), (V,f), (VI,a), and NB, were dissolved in isooctane which simulated the ethylene-propylene units present in EPTMs or EPDMs[25]. The approach was essentially that described by Szwarc et al.[50].

Ter-butoxy radical is known to be very reactive towards allylic hydrogen abstraction Eq. (2)[54], but a fraction undergoes the β scission reaction (1):

$$(CH_3)_3CO \cdot \xrightarrow{k_1} CH_3^{\cdot} + CH_3COCH_3 \tag{1}$$

$$(CH_3)_3CO \cdot + \text{Isooctane} \xrightarrow{k_2} (CH_3)_3COH + \text{Isooctyl radical} \tag{2}$$

In the presence of olefins (T) two other reactions have to be taken into account:

$$(CH_3)_3CO \cdot + T \xrightarrow{k_3} (CH_3)_3COH + T^{\cdot} \text{ (hydrogen abstraction)} \tag{3}$$

$$(CH_3)_3CO \cdot + T \xrightarrow{k_4} (CH_3)_3COT^{\cdot} \text{ (double bond addition)} \tag{4}$$

The kinetic scheme was verified during the course of the same work[25] and the reactivity of t-butoxy radicals toward the saturated substrate (k_2) and model compounds (k_3, k_4) was related to the β scission reaction (k_1). The results are listed in Table 7 which shows that the H abstraction mechanism from model compound (III,a) with respect to isooctane (k_3/k_2) is four times more favored than in the case of NB. The addition reaction of t-butoxy radical to unsaturation (k_3/k_4) is 15 times

Table 7. Relative rate constants for the reactions of ter-butoxy radicals with iso-octane and model compounds simulating different conjugated dienic unsaturations[a]

Model compound	$\dfrac{k_2}{k_1}$	$\dfrac{k_3}{k_1}$	$\dfrac{k_3}{k_2}$	$\dfrac{k_4}{k_1}$	$\dfrac{k_3}{k_4}$
Isooctane	0.36	–	–	–	–
(III,a)	–	31	88	< 2	> 15
(V,f)[b]	–	10 ± 2	29 ± 6	< 2	> 5
(VI,a)	–	2 ± 1.0	4 ± 2	14 ± 8	0.14
NB	–	8 ± 1.0	21 ± 1.0	< 0.5	> 16

[a] At 115 °C; see the text for the meaning of the rate constants.
[b] Extrapolated to zero concentration of model compound.

less important than the H abstraction in the case of (III,a) and NB; this order is reversed in the case of (VI,a). The latter result is surprising since the abstraction of one bisallylic hydrogen of (VI,a) yields a radical which is stabilized by conjugation with two double bonds and the entire system is coplanar[55, 56].

However, it is worth noting the ability of t-butoxy radicals to attack the saturated molecule of isooctane (k_2/k_1). This has also been recently confirmed in the case of linear saturated hydrocarbons[57]. In the same work[57] it was found that the benzoyloxy radical disproportionation into phenyl radical and CO_2 is, at 140 °C, about four times more favored than the H abstraction from n-pentadecane, while in the case of t-butoxy radical the H abstraction is higher than the β scission reaction by a factor of eight.

These data outline the great importance of free radicals involved in grafting processes and indicate the possibility of grafting saturated backbones as well, although with reduced efficiency.

The same approach has been recently adopted[95] for studying the behavior of (III)-EPTM in the presence of benzoylperoxide. The reaction scheme firstly considered was:

a) $BZP = 2 \ PhCO_2^{\cdot}$
b) $PhCO_2^{\cdot} = Ph^{\cdot} + CO_2$
c) $PhCO_2^{\cdot} + (III,a) = PhCOOH + R_1^{\cdot}$
d) $PhCO_2^{\cdot} + (III,a) = R_2^{\cdot}$

By determining experimentally the yield of PhCOOH and CO_2 in the presence and in the absence of (III,a), it has been possible to find, by using the competition method[25], the following values (at 83 °C): $k_c = 2.7 \cdot 10^6$ and $k_d = 4.8 \cdot 10^6 \ 1 \cdot mole^{-1}$. sec^{-1}, referred to eq. c) and d), respectively.

However, such an evaluation may be complicated by induced decomposition of the peroxide and/or by interference of highly reactive products from the primary reactions. Experiments with different concentrations of (III,a) confirmed that these complications exist in our case and we were forced to extend our reaction scheme as follows:

e) $R_1^{\cdot} + R_2^{\cdot} = P$
f) $PhCO_2^{\cdot} + P = PhCOOH + P^{\cdot}$
g) $PhCO_2^{\cdot} + P = PhCO_2P^{\cdot}$

from which we obtained a more reliable result for $k_c + k_d = 5.8 \cdot 10^6 \ 1 \cdot mole^{-1} \cdot sec^{-1}$.

As far as the ethylene propylene moiety is concerned, we used benzene solutions of EPM and the reaction scheme comprising equations a)–d) put forward for (III,a) since no competition reaction is expected. We obtained at 83 °C: $k_{abstr(EPM)} = 8 \cdot 10^3 \ 1 \cdot mole^{-1} \cdot sec^{-1}$.

In another paper[97], the reactivity of methyl radical toward the unsaturation present in MNB-EPDM, ENB-EPDM, DCP-EPDM and (III)-EPTM was investigated. The model compounds of MNB, ENB, DCP and (III), i.e. 2-methylene-norbornane (MB), 2-ethylidine-norbornane (NB), tricyclo-[5.2.1.0.$^{2.6}$]-dec-8-ene (TDE) and diene (III,a), were subjected to react with diacetyl peroxide in isooctane according to the procedure above described for t-butoxy radical. The reaction scheme is given, in the present case, only by eqs. (2)–(4) and by substituting $(CH_3)_3CO^{\cdot}$ with $CH_3^{\cdot}$.

Table 8. Relative rate constants for the reactions of methyl radicals with isooctane and the model compounds simulating the unsaturation present in some ethylene-propylene based terpolymers

Model Compound[a]	$\dfrac{k_4}{k_2}$	$\dfrac{k_3}{k_2}$	$\dfrac{k_3}{k_4}$
(III,a)	55.6	8.0	0.15
MB	ca. 40	n. d.	n. d.
NB	2.5	1.4	0.56
TDE	2.4	2.9	1.21

[a] (III,a) = 9,10-dihydro-1-isopropylidene-dicyclopentadiene
   MB = 2-methylene-norbornane
   NB = 2-ethylidene-norbornane
   TDE = tricyclo-[5.2.1.0$^{2.6}$]-dec-8-ene, i. e. 9,10-dihydro-dicyclopentadiene

The results obtained are reported in Table 8 and show that the methyl affinities (k_4/k_2) of MB, NB and TDE are comparable with those known for acyclic and cyclic olefins, respectively, having approximately the same degree of substitution at the double bond. Conversely, the k_4/k_2 ratio for (III,a) is 5 times greater than that expected for two separated double bonds with comparable degree of substitution, but 1−2 orders of magnitude smaller than those observed for other dienic systems, e. g. butadiene, cyclopentadiene, etc. This result illustrates the opposite effects due to resonance stabilization and steric hindrance.

The k_3/k_2 ratios give the order of hydrogen abstraction reactivity, but the values obtained indicate that the tendency is much lower than expected for acyclic and strain free olefins having an equal number and type of α-hydrogen atoms. The results can be explained by attributing to the tertiary bridgehead sites adjacent to the double bonds a lack of reactivity due to the strain energy necessary to acquire the planar configuration of the radical.

Finally, a comparison of Table 8 and 7 shows that the tendency of $(CH_3)_3CH^{\cdot}$ radical to abstract hydrogen atoms from NB and (III,a) is, respectively, 30 and 100 times higher than that shown by $CH_3^{\cdot}$ radical.

c) Grafting of Isobutene-Triene Copolymers

As it will be discussed in Sect. III.B.2.d), isobutene homopolymer and isoprene-isobutene copolymer (butyl rubber) are prone to degrade under the attack of free radicals[58]. Therefore, these polymers cannot cross-link (unless high comonomer content is present) in the presence of free radicals and usually are not used in grafting reactions since a low degree of graft is observed (Table 9, n° 1,2).

　　　　　　　　　　　　　　　　　　　　　　　　S. Cesca, A. Priola, and M. Bruzzone

Table 9. Styrene-acrylonitrile grafting onto isobutene copolymers[a]

Run n°	Type[b]	Isobutene Copolymer Comonomer (mol %)	$[\eta]$ (dl/g)	Conv. (%)	Elastomer in final polymer (%)	Degree of graft[c] (%)	$[\eta]$ of SAN (dl/g)	Impact resistance[d] (J/m)	Ungrafted copolymer (%)
1	IIR	1.40	2.20	88	11.6	7.8	0.72	48	70
2	IIR	1.90	1.85	83	12.5	16.0	0.57	n. d.	51
3	HTI	1.00	1.65	83	12.5	36.6	0.49	n. d.	25
4	HTI	1.87	2.05	84	11.9	52.0	0.43	88	n. d.
5	OTI	1.90	1.52	86	11.8	18.7	0.73	65	47

[a] Conditions: T = 85 °C; initiator = BPO; other conditions as in Tables 3 and 4.
[b] IIR = Poly (isobutene-co-isoprene); HTI = Poly(isobutene-co-1,3,5-hexatriene); OTI = Poly(isobutene-co-2,4,6-octatriene).
[c] Defined as in Table 4.
[d] Notched Izod Test.

Interestingly, copolymers of isobutene with 1,3,5-hexatriene (HTI) show a noticeable increase in grafting efficiency with respect to butyl rubber when styrene and acrylonitrile are grafted by means of BPO (Table 9): the same result is not observed in the case of 2,4,6-octatriene based copolymers (OTI). These experimental facts suggest that the role played by the allylic methylene groups introduced in HTI chains is essential in allowing grafting processes occurring through a prevalent hydrogen abstraction mechanism (Fig. 9). When the allylic positions carry a methyl group, as in the case of OTI, the residual allylic hydrogen atoms apparently are shielded by adjacent methyl groups and, similar to butyl rubber, a low degree of graft is obtained. Therefore, the contribution to the grafting process induced by BPO and involving free radical addition to the *trans-trans* conjugated double bonds present in isobutene-triene copolymers is of secondary importance with respect to the hydrogen abstraction mechanism, as indicated also by model compound reactivity (Figs. 9 and 10).

The block distribution of 2,4,6-octatriene existing in OTI (Sect. III.A.2.) can be another reason accounting for the lower grafting efficiency of OTI with respect to HTI, since allylic positions are more sterically hindered in the triene blocks.

2. Radical and Sulfur Curing

The problem of cross-linking or grafting low unsaturated rubbers (e. g. EPDMs or EPTMs) or thermoplastics (e. g. ethylene-butadiene copolymer[59]) is fundamentally the same. Radical mechanisms are believed to be operative when peroxides or sulfur-based formulations are used, even though in the latter case ionic mechanisms also seem to contribute to the curing process[60].

While a lot of technological work has been done in the past to find the best applications for a number of new polymers, relatively little information has been obtained about the basic chemistry of cross-linking processes.

The unusual behavior of some terpolymers containing a system of conjugated double bonds which show a high rate of vulcanization and yield high cross-linking density, even when the unsaturation content is low, induced us to investigate the cross-linking mechanism of fast-curing EPTMs. Low molecular weight compounds simulating the unsaturations existing in EPTMs were used to study the modification of the reactive sites. Efforts were made to treat quantitatively the results obtained with both model compounds (Sect. B.2.a) and cured terpolymers (Sects. B.2.b and B.2.c) in order to elucidate the reaction mechanism of free radicals and the coexistence of an unusual curing process with the conventional one (Sects. B.3.).

a) Studies with Model Compounds

The model compounds (III,a), (V,f), (VI,a) and NB described in Sect. B.1.b) and the kinetic results obtained from the attack of t-butoxy radical on isooctane solutions of these products[25], have been used to clarify some aspects of the mechanism of terpolymer cross-linking. The reactivity of isooctane, expressed as the hydrogen abstraction reaction by t-butoxy radical relative to the β scission reaction of $(CH_3)_3CO^{\cdot}$ (see k_2/k_1 values in Table 7), has been converted into the rate constant for the H abstraction from ethylene-propylene units (k_5) by means of Pryor's

Table 10. Amount of t-butoxy radicals undergoing β scission in the
presence of different ethylene-propylene-based elastomers

Polymer (type)	Termonomer concentration (mole/l)	Upper limit of CH_3 radical formed (%)
EPM	0	20
ENB-EPDM	0.2	15
(III)-EPTM	0.2	10
(V)-EPTM	0.2	15
(VI)-EPTM	0.2	15

equation[61]. Thus $k_5/k_1 = 1.5 (k_2/k_1)$ has been obtained. The assumption that the
relative reactivities determined from experiments using solutions of elastomers are
not drastically influenced by the high viscosity of solid terpolymers, seems partly
justified by some recent results[57]. Since t-butoxy and methyl radicals are generated
by the decomposition of di-t-butyl-peroxide, and because they display great difference
of reactivity toward H abstraction and double bond addition[62], their relative concentration has been calculated by taking into account the composition of the substrate
(EPM, EPTM or EPDM) and the values of k_5/k_1 (Table 10). The results indicate that the
amount of $(CH_3)_3CO^\cdot$ subjected to the β scission is of minor importance, since it
does not exceed 20%. This superior value, which is in substantial agreement with the
data of other authors[57], was obtained in the case of the less reactive substrate
(EPM), while the fraction of $CH_3^\cdot$ radical is lower when the reactivity of the terpolymer unsaturation (expressed as k_3/k_2 in Table 7) increases.

The reactivity of EPTMs in radical curing, predicted by these calculations, has
been confirmed directly by cross-linking density measurements (Table 11) also
carried out on EPM and some EPDMs vulcanized with dicumyl peroxide at 145 °C.
In Fig. 11 it is evident that an EPDM containing 0.54 mol/l of ENB exhibits the same
cross-linking efficiency as an EPTM containing 0.05 mol/l of (III).

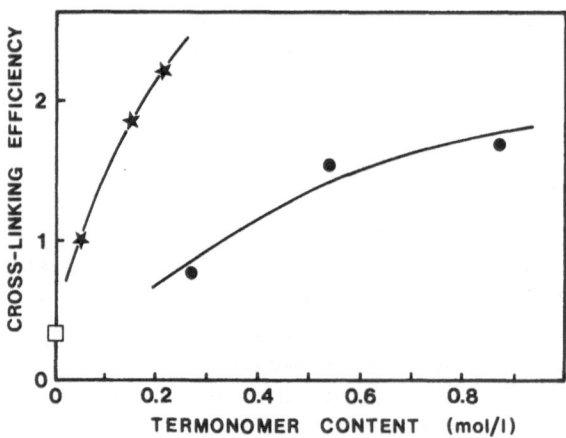

Fig. 11. Cross-linking efficiency of (III)-EPTM ($\star$) and ENB-EPDM ($\bullet$) as function of termonomer content. Conditions: dicumylperoxide = 1 wt %; T = 145 °C; time = 250 min

Table 11. Cross-linking density of some EPTMs, EPDMs and EPM cured with dicumyl peroxide (DCP)

Termonomer Type	Conc. (mol/l)	$\bar{M}_n \cdot 10^{-5}$	DCP (%)	$\nu_{tot} \cdot 10^4$ (mol/cm^3)	$\nu_c \cdot 10^4$ (mol/cm^3)	Cross-linking efficiency[c]
(III)	0.05	8.3	1-2	2.0-30	0.61-0.98	1-0.79
(III)	0.125	8.2	0.5-1-2	2.7-4.2-5.6	0.79-1.5-2.4	2.6-2.5-1.9
(V)	0.07	6.6	0.0-1-2	2.25-3.4-4.2	n. d.	n. d.
ENB	0.57	8.0	0.5-1-2	1.4-2.9-4.7	0.5-1.0-1.9	1.6-1.6-1.6
MNB[a]	0.46	8.4	0.5-1-2	5.1-6.3-6.7	2.1-2.9-3.2	7.0-4.7-2.6
—[b]	0	n. d.	1-2-4	0.2-1.4-2.9	0.2-0.5-1.0	0.34-0.39-0.40

Conditions: T = 145 °C; time = 250 min.; data obtained by swelling measurements in n-heptane at 30 °C according to G. Kraus, Rubber Chem. Technol. 30, 928 (1957).

[a] 5-methylene-2-norbornene.
[b] EPM.
[c] Number of chemical cross-links formed per mole of peroxide used.

Figure 12 shows the behavior of different terpolymers as a function of the unsaturation and the peroxide content in the formulation. The unusual trend of
(V)-EPTM can be explained by the existence of another cross-linking mechanism
(self-vulcanization) besides the radical one (see the next Section), while the high
yield of cross-links displayed by 5-methylene-2-norbornene-EPDM is due to the high
reactivity of the radical originating from the methylene double bond which undergoes, preferentially, addition reactions instead of coupling processes[63].

Another calculation was made[25] in order to distinguish between the amount of
t-butoxy radical reacting with ethylene-propylene units and that reacting with the
unsaturations of some terpolymers. The results (Table 12) indicate that there is always a significant amount of t-butoxy radicals which react with the H atoms of the
saturated back-bone, while in the case of (III)-EPTM 34% of $(CH_3)_3CO^{\cdot}$ reacts with
the unsaturation when the termonomer content is only 0.1 M. The same result is
achieved in the case of ENB-EPDM when the termonomer content is four-fold end
hence agrees with the experimental data of Fig. 11. However, the formation of
abundant amounts of tertiary alkyl type radicals, e.g. $\sim CH_2\text{-}\dot{C}(CH_3)\text{-}CH_2\sim$,
when the termonomer concentration is low, can explain the poor yield of grafting
and cross-linking exhibited by EPM or low unsaturated EPDMs. In fact, tertiary alkyl
radicals, which have been observed by ESR measurements in EPM at $-78\ ^{\circ}C$ after
γ-irradiation at 77 K[55], undergo a β scission and hence chain degradations occur,
while only secondary alkyl radicals, e.g. $\sim CH_2\text{-}\dot{C}H\text{-}CH_2\text{-}CH(CH_3)\sim$, are prone to
give coupling reactions[64] and hence effective cross-linking. However, the concentra-

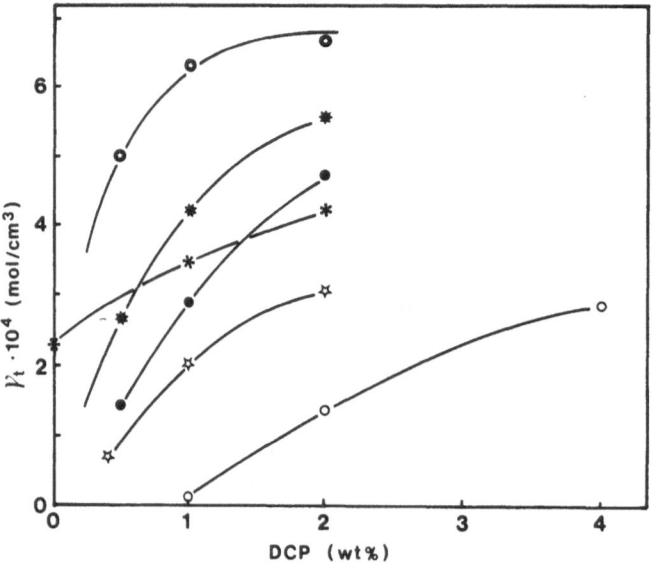

Fig. 12. Total cross-linking density obtained in different terpolymers as function of dicumyl-
peroxide (DCP) content. Conditions as in Fig. 11 (◐) = MNB-EPDM (MNB = 0.57 mol/l); (✳) =
(III)-EPTM ((III) = 0.13 mol/l); (●) = ENB-EPDM ((ENB) = 0.61 mol/l); (✱) = (V)-EPTM
((V) = 0.05 mol/l); (☆) = (III)-EPTM ((III) = 0.05 mol/l); (○) = EPM

Table 12. Fraction of t-butoxy radicals reacting with unsaturations of different ethylene-propylene-based terpolymers

Termonomer concn. (mole/l)	$(CH_3)_3CO^.$ radicals (%)			
	(III)-EPTM	(V)-EPTM	(VI)-EPTM	ENB-EPDM
0.05	20	9	12	6
0.10	34	16	20	10
0.20	50	25	35	19
0.50	71	46	58	38
1.00	83	63	73	57

tion of secondary alkyl radicals in EPM γ-irradiated at 77 K is relevant, but it becomes negligible at 195 K, where only tertiary radicals are evident, while at 223 K only ill-defined paramagnetic species (perhaps delocalized radicals arising from unsaturations eventually present in the main chain) can be observed[55]. Evidently, the lower energy of tertiary or allylic C—H bonds favors the observed transformation.

These experimental facts account for the low efficiency of EPM curing with peroxides. Furthermore, the decrease of MW (Tables 4 and 13) and the impossibility of grafting large amount of EPM in the presence of $(CH_3)_3CO^.$ or other free radicals (Table 14) may be explained by chain degradation phenomena subsequent to t-alkyl radicals attack. At the same time, preferred grafting of longer chains may occur[88, 89] and thus the findings of Tables 4, 13, 14 may have another explanation. As far as the reactive positions of termonomers are concerned, some interesting indications have been obtained by ESR investigations and HMO SCF calculations[25, 55, 56]. They were carried out on single γ-irradiated termonomers or their model compounds (eventually clathrated in thiourea or adamantane in order to restrict translational diffusion of the radicals and reduce anisotropic broadening of ESR patterns) and also γ-irradiated terpolymers.

According to the results obtained, mainly allyl and pentadienyl radicals are generated at room remperature or above, whereas at 77 K only alkyl radicals are formed in terpolymers. The latter arise through loss of hydrogen atoms from the ethylene-propylene units, but their conversion into more stable species occurs only on warming to 200 K. Evidently, below the glass transition temperature of terpolymers, radical formation is mainly governed by the relative abundance of ethylene-propylene units which is greater by a factor of 10—15 than termonomer units.

The radical species reported in Table 15, which refer to terpolymer based on (I), (III), (VI) and ENB, are the most probable. In the case of (ENB,a), (V,g) and (VI,b), steric hindrance should allow as more reactive position 1 and 2, respectively.

A contribution to find the more reactive positions in the case of (I,b), (III,b) and (VI,c) may be given by unpaired spin density calculations (Table 16) which indicate position 1 and, to a lesser extent, positions 3,5 (I,b and III,b) and 3,4 (VI,c). The results of ozonolysis experiments, carried out by Baldwin et al.[65] on cross-linked ENB-EPDM and which indicate that 24% of the cross-links occurs through position 1 of ENB unsaturation, agree with the predictions deduced from Table 16.

Table 13. Grafting of styrene and acrylonitrile onto EPM[a]

Run n°	Catalyst conc. (mmole/l)	Conv. (%)	EPM in final polymer (%)	Degree of graft (%)	Grafting efficiency (%)	Ungrafted EPM (%)	$[\eta]$ of ungrafted EPM (dl/g)
1	2.31	52.6	15.4	25.5	4.8	80.2	1.54
2	4.62	75.6	11.2	23.6	4.2	67.0	1.32
3	7.71	82.4	10.4	24.1	2.8	46.0	n.d.
4	13.9	89.0	9.7	30.9	3.4	34.8	1.03
5	27.7	90.0	9.6	22.0	2.4	23.9	0.80

[a] Conditions: As in Table 4, except initiator = ter-butyl 5-ethyl-perhexanoate; EPM having C_2H_4 = 71 mole % and $[\eta]$ = 1.98 dl/g.

Table 14. Grafting of styrene and acrylonitrile onto EPM[a] in the presence of different free radical initiators[b]

Run n°	Initiator (type)	(mmole/l)	T (°C)	Conv. (%)	EPM in final polymer (%)	Grafted EPM (%)	[η] of ungrafted EPM (dl/g)	Degree of graft (%)	Grafting efficiency (%)	[η] of SAN (dl/g)
1	AIBN	5.7	85	78.2	14.1	< 5	2.35	7	1.1	0.52
2	BPO	6.8	85	78.3	14.1	28.6	1.42	11.4	1.9	0.52
3	TBPH	7.8	85	82.1	13.5	37.3	1.30	23.7	3.7	0.62
4	TBPB	6.9	120	78.2	14.1	66.0	0.87	33.3	5.5	0.37
5	DTBP	12.1	130	84.8	13.1	63.0	0.76	82.4	14.1	0.33

[a] C_2H_4 = 74 mole %; [η] = 2,14 dl/g.
[b] Conditions and abbrevations as in Table 4.

Table 15. Most probable radical species originated by the unsaturations of some ethylene-propylene-based terpolymers

Type of termonomer	(I)	(III)	(V,e)	(VI)	(VI)	(ENB)
Type of radical	(I,b)	(III,b)	(V,g)	(VI,b)	(VI,c)	(ENB,a)

Table 16. Molecular orbital calculation of π spin density distribution of radicals (I,b), (III,b) and (VI,c)[a]

(I,b) and (III,b)[b]   (VI,c)[b]   (VI,c)[c]

[a] cf. Table 15.
[b] HMO SCF(Mc Lahlan) calculation.
[c] INDO restricted wave function calculation.

As far as the sulfur curing process is concerned, some interesting information was obtained by reacting the model compound (V,e) in decalin (12%) with sulfur (8%) at 145 °C in the dark for four hours and in an inert atmosphere. Mass spectrometric analysis of the volatile fraction showed that: (i) the molecule of the model compound added more than one sulfur atom; (ii) the Diels-Alder dimer of (V,e) was present (see Sect. B.3.a and the next Section); (iii) one sulfur atom linked two molecules of (V,e); (iv) degradation reactions yielded 1,2-dinorbornane-ethane and the addition of both one sulfur atom and one cyclopentadienyl group to (V,e). These results were confirmed by the compositional analysis of the network obtained in the vulcanization of (V)-EPTM (Sect. B.2.b)) and indicate the possibility of a Diels-Alder reaction between the cyclopentadienyl rings of two molecules of (V). On the other hand, model compounds of polyisoprene were found capable of adding several atoms of sulfur[66] as expected in conventional sulfur curing of polymers containing isolated double bonds.

b) Radical and Sulfur Covulcanization of EPTMs with Highly Unsaturated Elastomers

Since blends of ethylene-propylene-based terpolymers with highly unsaturated rubbers are of great interest because of their superior ozone resistance[65], we have investigated the possibility of utilizing the high reactivity of EPTMs to obtain true covulcanized blends. It is known, in fact, that blends of polydienes and polyolefin elastomers are heterogeneous dispersions of a fast-curing polydiene phase and a slow-curing EPDM phase and hence poor properties are exhibited by the vulcanizates which, in reality, consist of overcured polydiene and undercured EPDM mixture.

EPTMs are sufficiently reactive toward free radicals (see previous Section), even when they contain low amounts of conjugated unsaturation (0.1 M or less). Thus degradation phenomena are practically avoided during their radical curing. We have verified the possibility of co-curing them with cis-1,4-polyisoprene (IR) by means of dicumyl peroxide-based formulations. Figure 13 shows that with 0.5−1.5 % peroxide

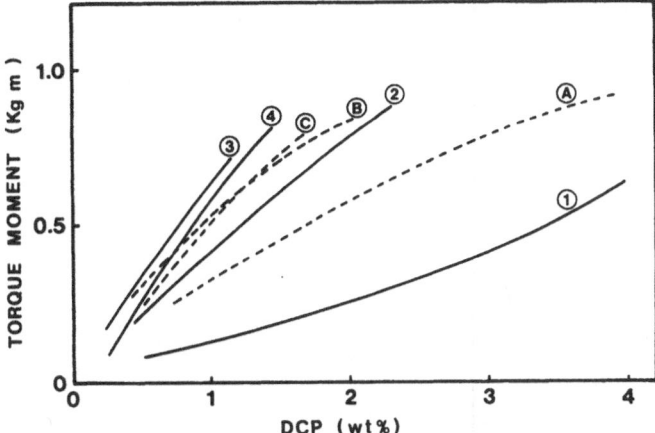

Fig. 13. Radical vulcanization of some terpolymers alone (continuous line) or blended (50 : 50; dotted line) with cis-1,40polyisoprene (IR) : dependence of the torque moment on the dicumyl peroxide content. Conditions as in Fig. 11: 1 and A = EPM; 2 = IR; 3 and B = EPDM (ENB = 0.61 mol/l); 4 and C = EPTM ((III) = 0.05 mol/l)

1:1 blends based on IR and EPDM containing 0.61 mol/l of ENB or EPTM containing 0.05 mol/l of (III) yield very close torque values. Tensile properties measurements and extraction data on vulcanizates confirmed that effective networks were obtained in the case of blends based on terpolymers, whereas EPM did not succeed in giving the same result even when amounts of peroxide greater than 2 % were used.

However, these interesting results could not be extended to the most usual sulfur vulcanization, since the mechanism of this curing process is very complicated and not yet well understood[65]. Experiments with typical sulfur formulations have shown (Figs. 14–19 and Ref.[9]) that both SBR and IR can covulcanize with EPTMs

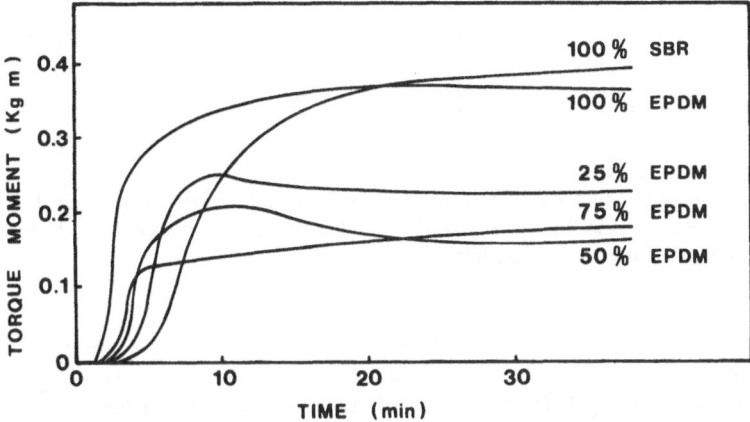

Fig. 14. Vulcanization kinetics of SBR, (II)-EPTM containing 0.40 mol/l of (II) and their blends. Recipe: polymer = 100; HAF = 50; Naphthenic oil = 5; ZnO = 5; TMTD = 1.5; S = 1.5; MBT = 0.5; T = 145 °C

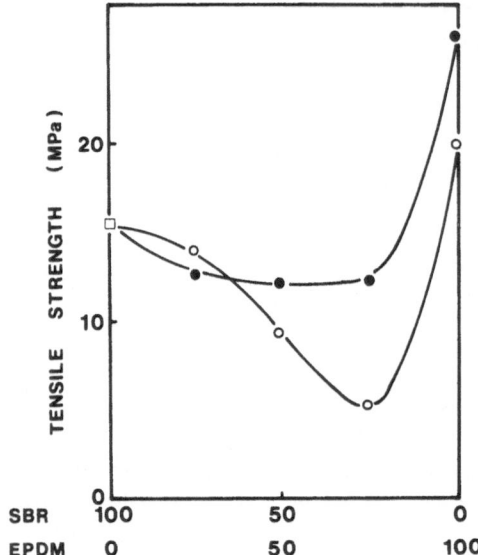

Fig. 15. Covulcanization of (II)-EPTM (•) (Sample of Fig. 14) with SBR (reference to ENB-EPDM containing 0.54 mol/l of diene (○)). Recipe as in Fig. 14; vulc. time = 60 min

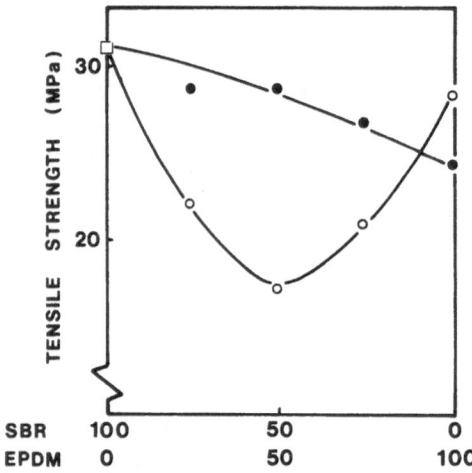

Fig. 16. Covulcanization of (V)-EPTM (•) (0.27 mol/l of triene) with SBR (reference to ENB-EPDM (○) of Fig. 15). Recipe as in Fig. 14, except: S = 1.7 and CBS (N-cyclohexyl 1-2-benzo-thiazole sulphenamide) = 1; vulc. time = 150 min

based on (II) or (V), since the values of tensile properties are not sharply reduced by increasing amounts of terpolymer. In contrast, unsatisfactory data have been obtained with EPDMs containing ENB or 1,4-hexadiene (termonomer concentration up to about 1 M) and also with EPTMs based on (III), (VII) and VIII). (VI)-EPTM showed a strong tendency to yield gel (self-vulcanization) which caused difficulties in covulcanization experiments.

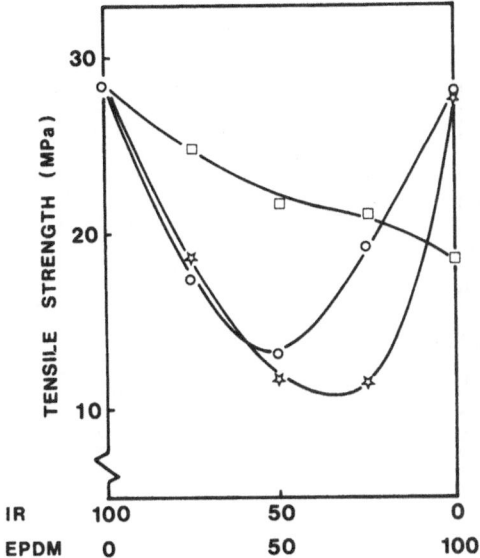

Fig. 17. Covulcanization of *cis*-1,4-polysoprene (IR) with (V)-EPTM (0.105 mol/l of triene) or with EPDMs based on 1,4-hexadiene (0.66 mol/l of diene) or ENB (0.92 mol/l). EPTM = (□); 1,4-HD-EPDM = (☆); ENB-EPDM = (○). Recipe: polymer = 100; HAF = 50; Circosol-4220 = 5; ZnO = 5; S = 1.7; CBS = 1; AO-2246 = 1; Stearic acid = 1; T = 145 °C; time = 90 min

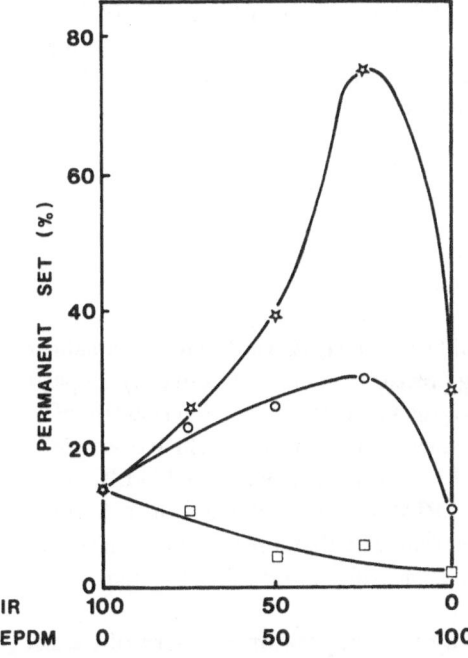

Fig. 18. Covulcanization of IR with some terpolymers. Sample, symbols and conditions as in Fig. 17

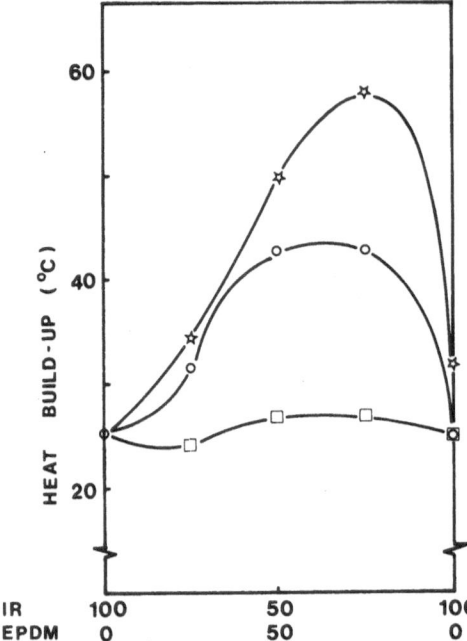

| IR | 100 | 50 | 100 |
| EPDM | 0 | 50 | 0 |

Fig. 19. Heat build-up of samples covulcanized according to Fig. 17

The interesting behavior shown mainly by (V)-EPTM was not confirmed by terpolymers containing 5,6-dimethylene-2-norbornene (X)[19, 94] or 5-(2'-vinyl-3'-butenyl)-2-norbornene (XI), both being model compounds of triene (V). In fact, the first EPTM contains two conjugated double bonds in obligatory cisoid conformation which were found[52] very reactive toward methyl free radicals. The second EPTM has one bis-allylic hydrogen atom in the unsaturated system:

(X)   (XI)

Therefore, the exceptional behavior of (V)- or (VI)-EPTM has to be explained by the reactivity of the cyclopentadienyl ring present in trienes (V) and (VI). In particular, the presence of two bis-allylic hydrogens seems to be the preferred site of reaction during the sulfur curing process, since the isomer (V,c), which does not carry allylic hydrogen at all, does not yield a fast-curing, covulcanizable EPTM. On the other hand, the ability of (V)- or (VI)-EPTM to cross-link under the action of the temperature alone (Fig. 12 and Sect. B.3.a)) suggests that a second mechanism is operative originating, very likely, from the Diels-Alder interchain condensation of cyclopentadienyl rings.

This conclusion is strongly supported by the compositional analysis of the network resulting from sulfur curing of two different terpolymers. In fact, the data obtained (Table 17) in the case of (V)-EPTM show:

Table 17. Network chain density data obtained at different curing times for (V)-EPTM and ENB-EPDM

Type of polymer	Vulcanization[a] time (minutes)	$\nu_{tot} \cdot 10^4$ [b] (mol/cm³)	$\nu_{c,tot} \cdot 10^4$ [b] (mol/cm³)	$\nu_n \cdot 10^4$ [b] (mol/cm³)	$\nu_2 \cdot 10^4$b [b] (mol/cm³)	$\nu_1 \cdot 10^4$ [b] (mol/cm³)
(V)-EPTM						
(V) = 0.39 mol/l	5	0.3013	0.1355	0.0096	0.0005	0.1254
[η] = 2.40 dl/g, toluene, 30 °C	10	0.8143	0.2087	0.0423	0.0130	0.1524
C$_2$H$_4$ = 71 mole %	15	1.7434	0.3815	0.1118	0.0001	0.2697
Gel = 2.5 %	20	1.4887	0.3266	0.0506	0.0201	0.2560
	30	1.3308	0.2958	0.0249	0.0252	0.2457
	60	1.6142	0.3528	0.0000	0.0166	0.3362
ENB-EPDM						
ENB = 0.71 mol/l	35	0.7928	0.3211	0.3211	—	—
[η] = 1.72 dl/g	42	0.8372	0.3308	0.3308	—	—
C$_2$H$_4$ = 69 mole %	50	1.1234	0.3968	0.3968	—	—
Gel = 0.9 %	65	1.2282	0.4225	0.1957	0.2267	—
	90	1.3520	0.4540	0.1869	0.2671	—
	180	1.4880	0.4901	0.1089	0.1225	0.2587

[a] Recipe: Polymer = 100; ZnO = 5; Sulfur = 1.7; Stearic acid = 1; AO-2246 = 1; Vulkacit CZ–C = 1; T = 145 °C.
[b] ν_{tot} = total cross-link density from equilibrium stress-strain data;
$\nu_{c,tot}$ = chemical total cross-link density from swelling measurements;
ν_n = polysulfidic link density; ν_2 = disulfidic link density; monosulfidic and C–C link density = ν_1 (s. Experimental).

(i)   The prevalence of C-C and C-S-C links; they are present in the early stages of the
      vulcanization process together with polysulfidic links which are the first to be
      formed[67] and this fact might indicate that C-C links are prevalent with respect
      to C-S-C bonds.
(ii) The low concentration of $C-S_2-C$ links;
(iii) The absence of $C-S_n-C$ links at elevated times of curing;
(iv) The high value of the overall cross-link density (i. e. including trapped entangle-
      ments besides chemical cross-links) in relation with the low concentration of
      unsaturation in the initial elastomer.

This situation is reversed in the case of ENB-EPDM (Table 17) since, after high
vulcanization times, relevant fractions of polysulfidic and disulfidic links are still
present, while the overall cross-link density is rather low in relation with the unsaturation content.

Furthermore, the plot of modulus vs. cross-link density (Fig. 20) suggests the
existence of two different types of network in the case of (V)-EPTM and ENB-
EPDM. The network obtained from EPTM curing appears more rigid on the basis of
the tensile properties values and this evidence agrees with a possible cross-link
functionality higher than 4 as a consequence of the existence of conjugated unsaturations in EPTM[68].

Of course, the conclusion descending from the experimental facts above
mentioned does not apply to EPTMs based on (I)-(IV) since their exo system of
conjugated double bonds cannot behave as a diene in Diels-Alder reactions. The
introduction of a third methyl group in the transoid conjugated diene system of (I)
allows the corresponding (II)-EPTM to covulcanize with SBR (Fig. 15) even though
ca. 0.5 M of (II) is necessary. Indeed, an elevated number of allylic hydrogens (9 per
molecule) are present in (II) and hence the co-vulcanization of (II)-EPTM with highly
unsaturated elastomers should be ascribed to its high vulcanization rate constant
which compensates the low concentration of unsaturation, so that both the

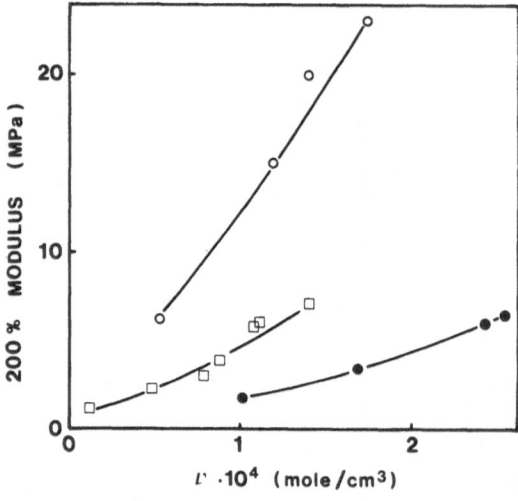

Fig. 20. 200% modulus vs. network
chain density for (V)-EPTM (o), ENB-
EPDM (=) and SBR (•): samples as in
Fig. 16

elastomers can yield a final interpenetrated network. This conclusion could be confirmed when an appropriate method could permit to know the relative reactivity of different terpolymers in vulcanization processes.

c) Kinetics of Vulcanization

A quantitative treatment of the complicated process of cross-linking in EPDMs has been proposed[13] on the basis of four main assumptions:

(i) Polysulfidic intermolecular cross-links are initially formed by the reaction of sulfur with unsaturation.

(ii) The break of cross-links formed originates active species.

(iii) Inactivation reactions involving the active specie (ii) imply the final destruction of cross-links.

(iv) The formation of stable interchain cross-links (very likely of monosulfidic type) is the consequence of reaction (ii).

The scheme summarized here is certainly a simplified representation of reality since it neglects the breaking of bonds involving primary chains and cyclization reactions. However, the overall vulcanization process can be described by two rate constants, i.e. k_i and k_f (Table 18), the former referring to the inactivation reactions and the latter to the reactions yielding the polymer network.

The final equations, descending from the elaboration of the kinetic expressions which describe the above reported assumptions, have been checked by evaluating some mechanical properties of the vulcanizates, because there are difficulties in determining the unsaturation content of the polymers along the curing time. An excellent concordance was found between experimental and calculated rate values, as it is shown in Figs. 21, 22 for two typical EPDMs and two EPTMs, respectively. The Arrhenius plot of k_f obtained at different temperatures for several EPDMs and (I)-EPTM[13] yields the same activation energy of 22 Kcal/mole and this common

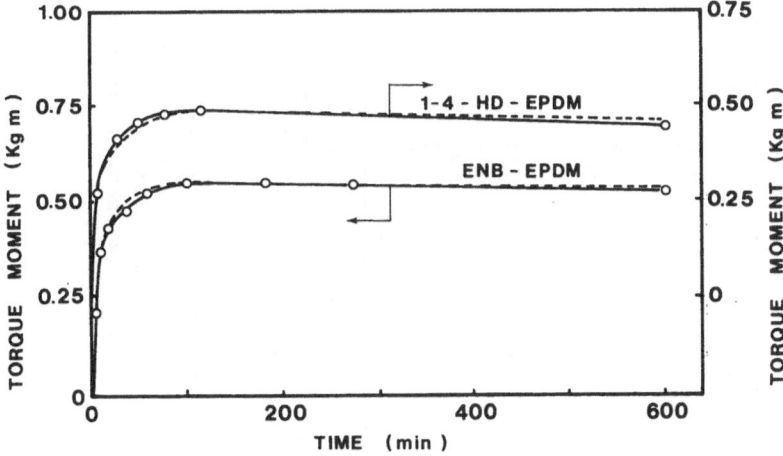

Fig. 21. Vulcanization kinetic curves of 1,4-HD (0.40 mol/l) and ENB (0.54 mol/l) based EPDM: full line = experimental curve; broken line = calculated curve (Ref.[13]). Conditions as in Fig. 14, except TMTD = 1, S = 2

Table 18. Vulcanization kinetic parameters of some unsaturated ethylene-propylene-based terpolymers[a]

Type[b]	Termonomer Concentration (mol/l)	k_f^c (kg^{-1}m^{-1}min^{-1})	$k_i \cdot 10^{3d}$ (min^{-1})	TM_{max}^h (kgm)	t_i^e (min)	t_{50}^f (min)	t_{90}^g (min)	$a/t_e 10^{2\,i}$ (min^{-1})
(I)	0.416	0.14	0.12	0.45	2.5	18	172	0.08
(II)	0.393	0.30	0.26	0.45	1.5	9	76	0.14
(V)	0.275	0.36	0.30	–	2.0	–	23	0.03
(VI)	0.079	–	–	0.31	1.3	4	9.3	–
(VII)	0.251	0.16	0.13	–	3.8	–	–	0.09
(VIII)	0.167	0.08	–	–	3.6	–	–	–
(X)	0.299	–	–	0.26	1.7	45	168	–
(XI)	0.247	–	–	0.23	2.8	30	139	–
ENB	0.542	0.14	0.26	0.60	5.0	18	175	0.04
1,4-HD	0.404	0.14	5.00	0.58	6.0	19	–	0.13
endo-DCP	0.378	0.07	0.20	0.58	7.5	33	450	0.10
MNB	0.404	0.12	–	0.64	5.0	–	–	–
exo-DCP	0.459	0.12	0.30	0.51	4.5	21	320	0.02

a Curing formulation: Polymer = 100; HAF Carbon Black = 50; Naphthenic Oil = 5; ZnO = 5; S = 2; TMTD = 1; MBT = 0.5; Temp. = 145 °C.
b S. Table 1; ENB = 5-ethylidene-2-norbornene; 1,4-HD = 1,4-hexadiene; DCP = dicyclopentadiene; MNB = 5-methylene-2-norbornene.
c Cross-linking reaction rate constant.
d Scission reaction rate constant.
e Induction time.
f Time requested to reach 50 % of TM_{max}.
g Time requested to reach 90 % of TM_{max}.
h Highest value of the torque moment.
i Constant connected to the equilibrium time between cross-linking and scission reaction.

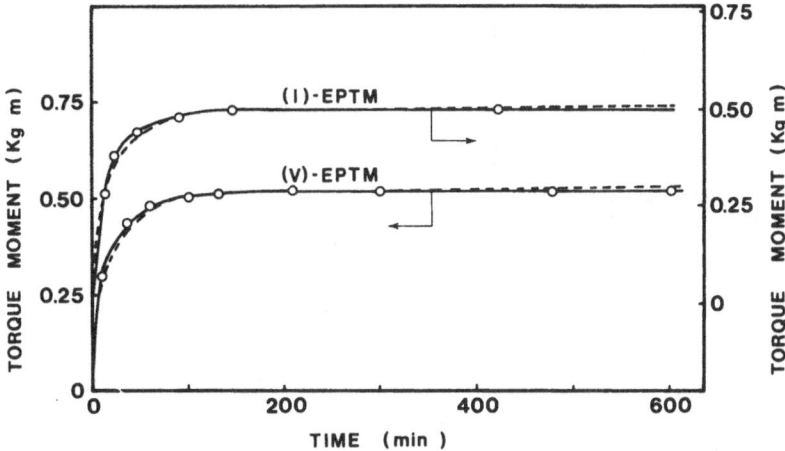

Fig. 22. Vulcanization kinetic curves of (I) and (V) based EPTM ((I) = 0.41 mol/l; (V) = 0.086 mol/l. Conditions and symbols as in Fig. 21

result may suggest that a common mechanism governs the curing process, at least when the unsaturation present in terpolymer is not prone to undergo side reactions (e. g. Diels-Alder condensation).

Aside from the possibility of distinguishing between two cross-linking mechanisms with other appropriate analyses (see previous Section), the data of Table 18 indicate that covulcanization processes occur, under the adopted conditions, when the cross-linking reaction rate constant of a given terpolymer is $\geqslant 0.3$ kg^{-1} m^{-1} min^{-1}. High rate constants imply, in principle, the possibility of obtaining high values of the tensile properties (Fig. 20), even when the unsaturation concentration is low.

d) Radical and Sulfur Curing of Isobutene-Trienes Copolymers

Previous investigations have shown that polyisobutene and poly(isobutene-*co*-isoprene), i. e. butyl rubber, are unable to cross-link in the presence of free radicals, since extensive chain scissions occur and thus low molecular weight products are formed[58]. The degradation mechanism proposed by Loan[58] involves, in the case of polyisobutene, the H abstraction from methyl groups followed by chain scission. Apparently, the formation of secondary alkyl radicals, which are believed to be responsible of polyolefin radical curing[64], is prevented for steric reasons by the presence of two adjacent dimethyl substituted carbon atoms and hence β scission reactions prevail.

This situation is substantially unchanged in the case of butyl rubber, where a few percent (0.6–3 mol %) of *trans*-1,4-isoprene units, i. e. C–CH$_2$–CH=C–CH$_2$–CH$_2$–C, are present and each methylene group is adjacent to at least one methyl substituted carbon atom. In fact, Loan has shown that the unsaturations of butyl rubber react more easily with the fragments arising from the dicumyl peroxide decomposition than the H atoms of isobutene methyl groups. The ratio of the two rate constants is

Table 19. Radical curing of isobutene copolymers[a]

Run n°	Isobutene Copolymer[b]		[η] (dl/g)	Gelled rubber[c] (%)	Tensile Properties[d]			
	Type	Comonomer (mol %)			100 % M (MPa)	300 % M (MPa)	TS (MPa)	EB (%)
1	HTI	1.10	1.90	61	0.5	0.9	3.3	900
2	HTI	1.73	n.d.	65	0.7	1.5	4.4	870
3	HTI	1.80	2.23	79	0.8	2.1	7.0	670
4	HTI	2.16	2.76	83	2.8	–	12.3	250
5	HTI	2.51	1.95	94	3.6	–	13.3	255
6	OTI	2.90	2.21	71	n.d.	n.d.	n.d.	n.d.
7	OTI	4.30	2.03	77	1.0	3.3	6.9	570
8	IIR	2.35	1.90	20	e	e	e	e

[a] Conditions: Dicumyl peroxide = 3 phr; T = 153 °C; time = 60 .

[b] Abbreviations as in Table 6.

[c] Residue to extraction with boiling cyclohexane (24 hours).

[d] M = modulus; TS = tensile strength; EB = elongation at break.

[e] Degradated sample.

about 300. Therefore, when the isoprene content is relatively high, e. g. 3 mol %, the degradation of butyl rubber is less pronounced and a small radical curing can be observed.

Attempts were made to overcome the difficulties above mentioned by ter-polymerizing isobutene, isoprene and divinylbenzene, to partially cross-linked products, which can give further radical curing[69].

The presence in polyisobutene chains of conjugated unsaturations highly reactive toward free radicals, as it occurs in the case of isobutene-triene copolymers, induced us to investigate radical cross-linking of these copolymers.

The results obtained, when dicumyl peroxide is used as curing agent (Table 19), indicate that 1,3,5-hexatriene (HTI) and 2,4,6-octatriene (OTI)-based copolymers can be easily cross-linked.

In the case of HTI at least 1.8 mol % of triene is necessary to obtain 80% of gelled rubber and hence good tensile properties of the vulcanizate.

The lower reactivity of OTI with respect to HTI (Table 19) is parallel to its behavior in radical grafting (Sect. B.1.c)), but in comparison with butyl rubber, OTI is more reactive, as far as the amount of gelled rubber indicates. However, levels of triene higher than 4 mol % are necessary to reach fair properties of the vulcanizate.

As mentioned in Sect. B.1.c) two main features differentiate HTI and OTI; (i) the block structure and (ii) the lower number of allylic hydrogens in OTI, which can explain the lower efficiency in radical reactions shown by the latter copolymer.

Attempts to elucidate the complex mechanism of sulfur curing of elastomers containing conjugated unsaturations were performed by studying the vulcanization kinetics of HTI and OTI (Table 20) containing different levels of triene on the basis of the approach described in Sect. B.2.c). According to Table 20, OTI appears faster curing than HTI and IIR, the latter copolymers exhibiting a similar behavior. But the

Table 20. Curing rate of sulphur vulcanized isobutene copolymers[a]

Comonomer		k^b	$t_{90}{}^c$
Type	Conc. (mol %)	$(kg^{-1}m^{-1}min^{-1})$	(min)
Isoprene	1.46	67	73
Isoprene	2.35	66	78
HT	1.44	88	77
HT	2.01	92	76
OT	1.97	119	63
OT	2.71	128	51

[a] Formulation: Polymer = 100; HAF = 50; ZnO = 5; stearic acid = 3; AO 2246 = 1; TMTD = 1; MBT = 0.5; S = 2; T = 153 °C.
[b] Curing rate constant according to Ref. [13].
[c] Time requested to reach 90 % of TM_{max}.

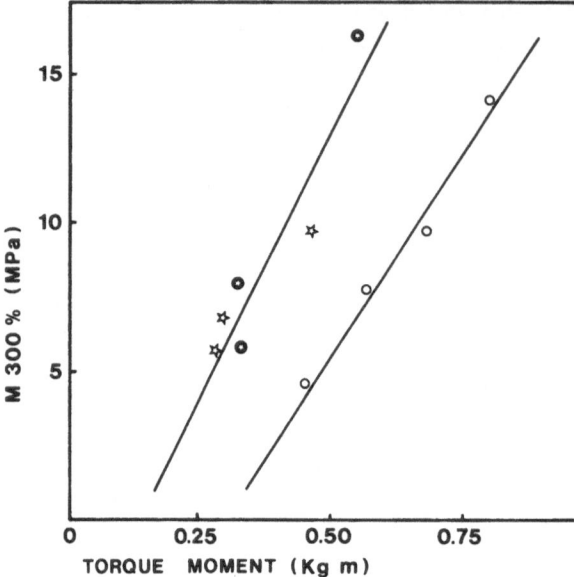

Fig. 23. 300 % Modulus vs. degree of cross-linking of some isobutene copolymers: (○) = IIR; (●) = HTI; (☆) = OTI; Recipe: polymer = 100; HAF = 50; ZnO = 5; Stearic acid = 3; Sulfur = 2; AO-2246 = 1; TMTD = 1; MBT = 0,5; T = 153 °C; time = 60 min

real meaning of the data pertaining to OTI is difficult to understand because of the structural complexity of this copolymer as outlined in Sect. A.2. However, Fig. 23 shows that the values of 300% modulus corresponding to a given value of the torque moment, i. e. a measure of the cross-linking density[70], are always higher when they refer to isobutene-trienes copolymers with respect to IIR. This experimental fact was also observed in sulfur-cured EPTMs and appears as a common feature of elastomers containing conjugated unsaturations. It can be attributable to the formation of a rigid network based on prevalent C-C links[68] (Sect. B.2.b)). Two sources of C-C bonds formation are likely: (i) Diels-Alder homocondensation reactions, which are believed responsible of the self-curing of HTI and also of butyl rubber containing conjugated unsaturation[8]; (ii) free radical reactions promoted by TMTD[71], which have been found operative in radical curing of HTI and OTI (see above).

3. Diels-Alder Reactions

While a system of two conjugated double bonds having an obligatory "transoid" conformation [(I)-(IV) trienes] is unable to form Diels-Alder adducts with dienophiles, a "cisoid" system (e. g. cyclopentadiene derivatives) reacts very easily[72]. This potential reactivity, derived from the ambivalent diene-dienophile behavior of conjugated diene systems capable of assuming the "cisoid" conformation[36], exists in EPTMs based on trienes (V)-(VIII). Since this possibility of reaction could contribute to the overall cross-linking process, we have investigated the possibility of Diels-Alder reactions both in model compounds and terpolymers. Of course, isobutene copolymers which contain *trans-trans* conjugated double bonds can also react according to the "cisoid" conformation and hence are expected to undergo Diels-Alder reactions easily.

In principle two modes of reactions are governed by the Diels-Alder mechanism, provided that sufficiently reactive diene systems are available: (i) homo-condensation of two diene groups, i. e. one diene group behaves as dienophile, (ii) condensation of a diene group with a dienophile. To the latter mode belongs the condensation of two conjugated diene functions with a bifunctional dienophile which can create a cross-link between the two chains carrying the conjugated unsaturation (the same final result can be derived from the reactions mode (i)).

a) Kinetics of Homo-Condensation of Cyclopentadienyl Model Compounds

Model compounds of triene (V), i. e. diene (V,f) or methyl-ethyl-cyclopentadiene (as in the case of (V), a mixture of positional isomers are obtained from the synthesis reaction), were found to undergo Diels-Alder reactions to appreciable extents even in the absence of dienophiles. The exclusive formation of dimers was checked by VPC-MS and the reaction course followed spectrophotometrically and through VPC. Since the reactions obeyed a second order kinetics at low conversions, we have extrapolated our experimental data to zero time[73–76].

Attempts to take into account the retro-Diels-Alder reaction in the treatment of the experimental data did not give a constant kinetic order as observed in the case of the reaction of (V,f) with 1,4-naphthoquinone (see below). On the other hand, the dissociation of cyclopentadienedimeric derivatives is practically negligible in the range of temperature explored, i. e. 50–90 °C. This conclusion is indirectly supported by the absence of information in the literature about this possibility. The results obtained are reported in Table 21 and are plotted in Fig. 24 according to an Arrhenius diagram in the case of (V,f). The value of $k_o = 2.0 \cdot 10^6 \, e^{-15,300/RT}$ l $\cdot$ mole$^{-1} \cdot$ min^{-1}, obtained between 50 and 90 °C, is comparable with the analogous results reported by other Authors: in the case of cyclopentadiene, dimerized in

Table 21. Kinetics of Diels-Alder dimerization of model compounds of triene (V)

Model compound	T (°C)	$k \cdot 10^4$ (l $\cdot$ mol$^{-1} \cdot$ min^{-1})
MEC[a]	70	3.01
(V,f)[b]	50	0.33
(V,f)	60	0.70
(V,f)	70	1.09
(V,f)	90	4.70

[a] Methyl-ethyl-cyclopentadiene, in hexadecane solution.
[b] In n-hexane solution.

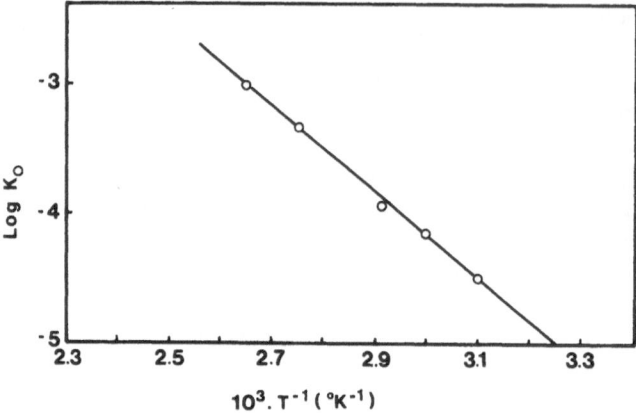

Fig. 24. Arrhenius plot of model compound (V,f) dimerization reaction

paraffin oil $k = 7.6 \cdot 10^8 \, e^{-17,400/RT}$, in benzene $k = 7.6 \cdot 10^7 \, e^{-16,400/RT}$, or in bulk $k = 3.8 \cdot 10^7 \, e^{-16,200/RT} \, l \cdot mole^{-1} \cdot min^{-1}$ [77]. The lower value of the frequency factor obtained for k_o reflects, very likely, the steric hindrance existing in the case of (V), while the activation energy (15,300 ± 12 cal/mole) is somewhat lower than in the case of cyclopentadiene[77]. Of course, k_o is an overall rate constant since it refers to the mixture of five positional somers which actually form triene (V).

The results of Table 21 indicate that, despite the steric constraint existing on the diene system of (V), fast Diels-Alder dimerizations occur with model compounds of (V) under mild conditions. The influence of highly viscous media on the course of bimolecular reactions is practically unknown, even though preliminary information which appeared recently[57], indicates that the influence of the solvent viscosity is scarce. But, in principle, the data of Table 21 permits us to ascribe the thermal vulcanization of (V)-EPTM (Fig. 12) to condensation reactions occurring between cyclopentadienyl moieties of different chains. These reactions, being mainly influenced by the temperature and the concentration of dienic unsaturations, appear as a cross-linking process concurrent with sulfur vulcanization and yield a network based on C-C links (Sect. B.2.b) and Table 17). Peculiar properties can be predicted for a vulcanized elastomer containing a noticeable number of C-C bonds[68], e. g. better behavior at high temperature and high values of the tensile properties (but also low elongation at break), and actually this behavior has been observed in vulcanized and covulcanized samples of (V)-EPTM.

b) Kinetics of Diels-Alder Adducts Formation Between Cyclopentadienyl Model Compounds and 1,4-Naphthoquinone

When one takes into account Diels-Alder reactions of cyclopentadiene derivatives with dienophiles, the concurrent reaction of dimerization of the diene has to be considered. Furthermore, it is known that Diel-Alder adducts can undergo dissociation reactions and hence the kinetic scheme must involve also the retro-Diels-Alder reactions. However, the dimerization reaction of (V,f) was $4 \div 5$ orders of magnitude lower than the addition reaction to 1,4-naphthoquinone (NQ) and hence the former has been neglected in elaborating our experimental data. This assumption has been supported by an iterative procedure (see Experimental Data) carried out by means of a computer, which showed that the concentration of dimers of (V,f) was practically negligible over a long period of time.

In order to avoid some difficulties during the treatment of the experimental data (change of the kinetic order when the conversion increases; difficulties of preparing a pure adduct between (V,f) and NQ), we have followed the approach of Wasserman[72], who deduced the value of k_2 from $K_{eq} = k_1/k_2$ (see Experimental).

The values of k_1, k_2 and K_{eq} obtained for the reactions of (V,f) and also (VI,a) with NQ, respectively at 62 and 80 °C, are listed in Table 22. These data show that NQ reacts faster with (V,f) than (VI,a) (the second reaction occurs at an appreciable rate only above 80 °C), but the latter adduct is more stable.

The Arrhenius diagrams give for the rate constants pertaining to the reaction of (V,f) with NQ the values reported in Table 23. The comparison of our kinetic data with those available for the cyclopentadiene-NQ adduct[78, 79], shows that (V,f)-NQ adduct is less stable by 9.1 Kcal/mole, but its formation activation energy is lower by 5.7 Kcal/mole. Also the frequency factor is lower in the case of (V,f)-NQ formation, and all the data suggest that the steric hindrance existing on the cyclopentadienyl ring of (V,f) exerts its influence on the dissociation reaction rather than on the adduct formation.

Table 22. Kinetic data of Diels-Alder reactions[a] between 1,4-naphthoquinone (NQ) and model compounds of trienes (V) and (VI)

(V,f) + NQ at 62 °C:	$k_1 = 4.9 \ 1 \cdot mol^{-1} \cdot min^{-1}$
	$k_2 = 0.98 \cdot 10^{-2} \ min^{-1}$
	$K_e = 500 \ mol^{-1} \cdot 1$
(VI,a) + NQ at 80 °C:	$k_1 = 8.9 \cdot 10^{-2} \ 1 \cdot mol^{-1} \cdot min^{-1}$
	$k_2 = 1.1 \cdot 10^{-5} \ min^{-1}$
	$K_e = 8190 \ mol^{-1} \cdot 1$

[a] In toluene solution.

Table 23. Dependence of the rate constants pertaining to the Diels-Alder reaction between NQ and (V,f) on the temperature[a]

$k_1 = 3.99 \cdot 10^4 \, e^{-5,900/RT} \ 1 \cdot mol^{-1} \cdot min^{-1}$
$k_2 = 3.36 \cdot 10^{11} e^{-19,900/RT} \ min^{-1}$
$K_e = 1.19 \cdot 10^{-7} e^{14,000/RT} \ 1 \cdot mol^{-1}$

[a] In toluene solution, between 23 and 62 °C.

c) Post-Modification of EPTMs

The modification of polymers by means of Diels-Alder reactions has been previously described in the case of 1,3,5-hexatriene, 1,3,5-heptatriene and 2,4,6-octatriene homopolymers[80]. Despite the *trans-trans*-1,3-diene structure of the repeat units of 1,6-polytrienes, which is known to react with dienophiles at a rate five times that of *cis-trans* dienes, the reactions were largely incomplete even after long periods of time when they were carried out at moderate temperature with strong dienophiles.

We have attempted to verify the indications emerging from the kinetic investigations carried out on model compounds (V,f) and (VI,a) also in the case of the corresponding EPTMs.

Since the influence of the polymer chains on the reactivity of functional groups present on them (due to steric and stereochemical effects, viscosity of the reaction medium, preferred solvation phenomena, influence of the neighbouring groups, etc.) and the equilibrium character of Diels-Alder reactions is known[81], an excess of dienophile was allowed to react with a dilute solution of EPTM in order to induce the formation of the largest quantity of Diels-Alder adduct possible.

Table 24. Reactions of (V)-EPTM with different dienophiles[a]

Conc. (V) (mol/l)	Dienophile type[b]	Mole ratio dienophile/ (V)	Reaction time (hr)	% of reacted[c] (V)
0.040	MVK	2.60	65	85
0.090	MVK	2.60	140	88
0.092	MVK	7.20	90	85
0.105	MVK	3.00	60	77
0.172	MVK	13.30	40	90
0.100	NQ	6.10	20	83
0.125	NQ	2.00	24	79
0.128	NQ	1.50	24	55
0.172	NQ	1.38	60	85
0.040	DAC	2.80	140	100
0.070	DAC	1.50	140	67
0.092	DAC	2.50	72	84
0.105	DAC	2.00	60	57
0.040	FN	1.09	65	90
0.090	FN	2.00	65	94
0.092	FN	1.40	70	94
0.105	FN	1.50	60	89
0.776	ACN	2.60	24	3
0.776	FMI	2.60	100	100

[a]   At 20 °C; 2% EPTM solution in toluene.
[b]   MVK = methylvinylketone; NQ = 1,4-naphthoquinone; DAC = dimethylacetylene dicarboxylate; FN = *trans*-1,2-dicyano-ethylene; ACN = acrylonitrile; FMI = N-phenylmaleinimide.
[c]   Obtained by measuring the intensity of the absorption at 252 nm due to the cyclopentadienyl system of (V). In the case of NQ the IR stretching band of carbonyl was used. Polymers were twice redissolved and reprecipitated from acetone before analysis.

The results reported in Table 24 indicate that the reactions can be complete or almost complete when strong dienophiles (e.g. N-phenylmaleinimide, fumaro nitrile of dimethylacetylene dicarboxylate) are used. However, the dienophiles listed in Table 24 are not the most reactive with cyclopentadiene[36] and hence long reaction times and an excess of dienophile is necessary to approach the reaction completeness.

The calculation of the amount of (V)-EPTM which should be transformed by reacting with NQ on the basis of the kinetic rate constant reported in the previous section, can give an idea of how much the presence of a macromolecular chain influences the reaction extent of a substituted cyclopentadienyl group involved in a Diels-Alder reaction. According to the data of Table 24 about 80% of (V), present in our EPTM containing 0.125 M of triene and dissolved (2%) in toluene, reacts with a twofold excess of NQ in 24 hours, whereas according to the kinetic rate constant of the model compound 95.0% of (V) should be transformed in 12.1 hours.

However, the possibility of reaching different degrees of completeness in Diels-Alder reactions involving EPTMs, depends mainly on the dienophilicity of the reagent. The order found was: N-phenyl-maleinimide > *trans*-1,2-dicyanoethylene > dimethylacetylene dicarboxylate > 1,4-naphthoquinone > methylvinylketone > acrylonitrile, in agreement with the reactivity order found for low molecular weight products[36].

Purified samples of (V)-EPTM heated in the dark and under inert atmosphere were found to contain increasing amounts of gel when the heating time increased (Fig. 25). By plotting the initial rate of gel formation vs. the reciprocal of the absolute temperature of heating, an apparent activation energy, E_{app}= 17.0 Kcal/mole, was obtained (Fig. 26). This value is in the range of the activation energies found (15.1−17.4 Kcal/mole) for the dimerization of cyclopentadiene performed in

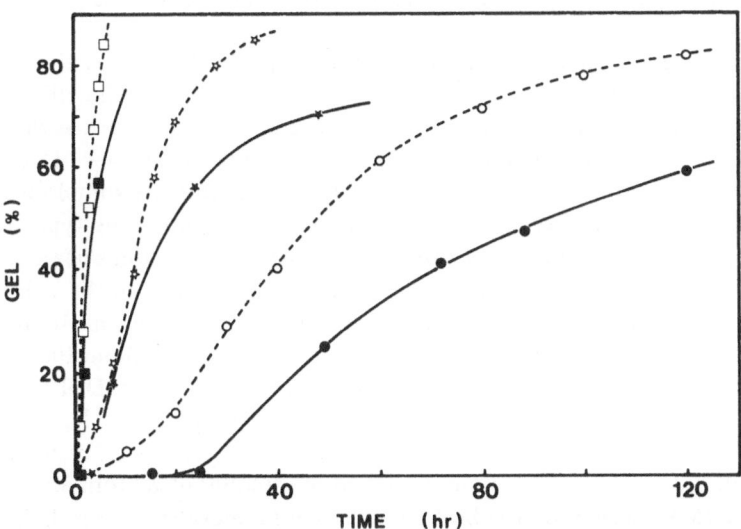

Fig. 25. Gel formed vs. time in (V)-EPTM samples heated at 50 (•), 70 (⋆) and 90 °C (■). Broken line = calculated values according to Langley's theory[82]

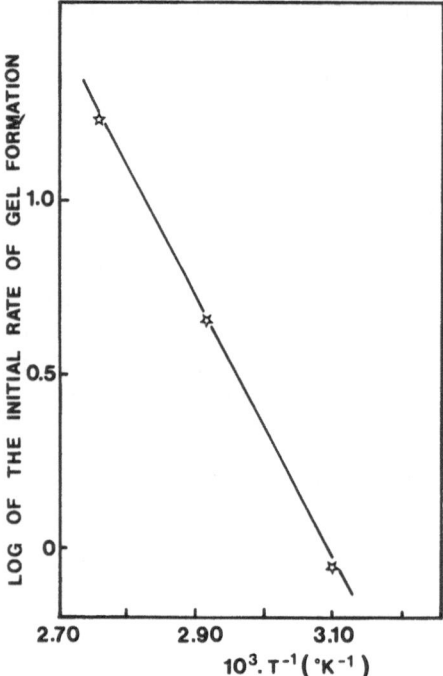

Fig. 26. Arrhenius plot for the initial rate of gel formation in (V)-EPTM ((V) = 0.319 mol/l)

different solvents[77]. The experimental conditions adopted and the value of E_{app} obtained from Fig. 26 clearly support the possibility of cross-links formation through Diels-Alder reactions in polymers having highly reactive diene groups, but also the possibility of relatively easy chain motion (see the T_g value of EPTMs[9]), in spite of the huge viscosity of the polymer mass ($\overline{M}_n \cong 10^5$) and the steric requirements for Diels-Alder reactions[36].

At present we have no kinetic data concerning the self-gelification of reactive terpolymers such as (V)-EPTM, and we cannot compare the behavior of a functional group inserted in a polymer chain with that present in a small molecule, such as the model compound (V,f). Such a comparison would be very interesting in view of the possibility of defining the reactivity of chemical functions present in macromolecules which are influenced by steric effects, viscosity of the reaction medium, and type of neighboring groups, among other things. However, systematic investigations of these effects have only recently begun[81]. One may be surprised, for instance, at the lack of any influence due to conformation in the chlorination of paraffinic chains. In this case we can extrapolate to the polymer level any information obtained about the reaction using small molecules. This situation seems unlikely in the case of Diels-Alder reactions involving EPTMs which contain cyclopentadienyl groups, at least on the basis of some calculations made with the kinetic data found with model compound (V,f) (Sect. B.3.a)). In fact, we have calculated the amount of gel formed in samples of (V)-EPTM subjected to mass heating under inert atmosphere in the dark, by assuming that the reactive sites of the polymer were involved in Diels-Alder reactions only. The calculation was made by means of Langley's statistical cross-linking theory which permits relating the number of reacted sites of a given polymer (provided that the molecular weight, the molecular weight distribution and the ratio between scissions and cross-links are known) to the gel content when a sufficiently high number of reactive sites per macromolecule are present[82]. Unfortunately, our sample of (V)-EPTM did not fully satisfy the last condition, since it contained a low amount of unsaturation. However, the comparison was attempted on the basis of an approximation between calculated and experimental gel values (Fig. 25) and it was evident that the calculated amounts of gel were always higher than the experimental ones, the difference being more noticeable at lower temperatures and at higher reaction times. Evidently, the possibility of reaction of the dienic system existing in the small molecule of model compound (V,f) is higher, even in the early stages, than in the terpolymer chain. This difference is enhanced when the possibility of motion of the macromolecules is depressed, i. e. at lower temperature or when the molecular weight of the chains increases.

d) Post-Modification of Isobutene-Trienes Copolymers

The more interesting isobutene copolymers described in Sect. A.2. contain as repeat units 1,4-di-substituted-1,3-dienes (Table 2). The common feature of these copolymers is the prevalent *trans-trans* configuration of the unsaturated system (infrared band at 985 cm^{-1}), because the infrared band at 970 cm^{-1}, due to the *cis-trans* dienic system, is very weak[3]. Since it is well known that Diels-Alder reactions of 1,4-*trans-trans* substituted derivatives of 1,3-butadiene are more favored than those of *cis-trans* isomers[83], we studied Diels-Alder reactions of the copolymers of Table 2 under mild and other conditions. Namely:

(i) In the absence of dienophiles, i. e. homocondensation reactions, from which dimers arise when low molecular weight products are employed, while in the case of polymer chains cross-links are formed (self-curing).

(ii) In the presence of monofunctional dienophiles, which introduce new chemical groups in the macromolecules by transforming the diene function into disubstituted cyclohexenic rings carrying anhydride, carboxylic, cyano, carbonyl, and other groups.

(iii) In the presence of bifunctional dienophiles which undergo reaction (ii) twice per each macromolecule and create a polymer network.

The previous work of Bell[80], carried out on homopolymer of 1,3,5-hexatriene, encouraged our expectations, since in the presence of different dienophiles the intensity of the polymer band at 985 cm^{-1} decreased markedly, whereas the absorption band at 970 cm^{-1} changed very little.

d 1) Diels-Alder Homocondensation. When isobutene-triene copolymers are heated to high temperatures, i. e. T $>$175 °C, relevant amounts of gel are formed. Figure 27 shows the increase in the mass viscosity of the polymer vs. the time of thermal treatment for HTI copolymer. An Arrhenius diagram was obtained by plotting the rate of gel formation vs. the reciprocal of the absolute temperature of heating (Fig. 28). The apparent activation energy of cross-links formation is E_{app}= 15.8 Kcal/mole, a value very close to the result found for butyl rubber containing conjugated double bonds

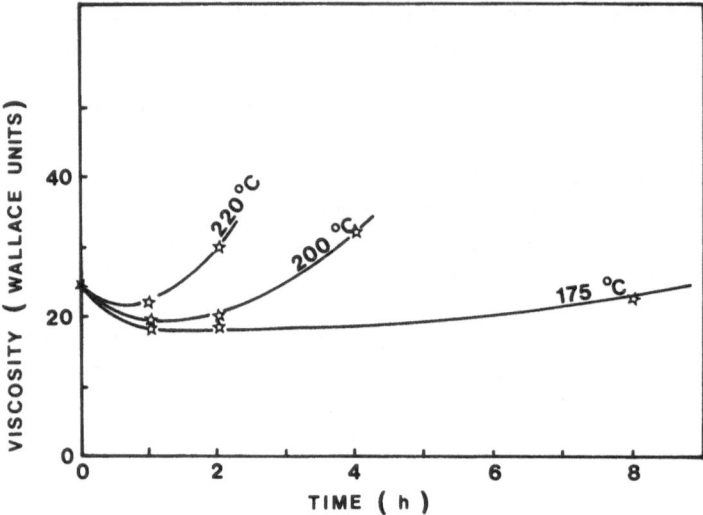

Fig. 27. Wallace viscosity vs. time of HTI at different temperatures

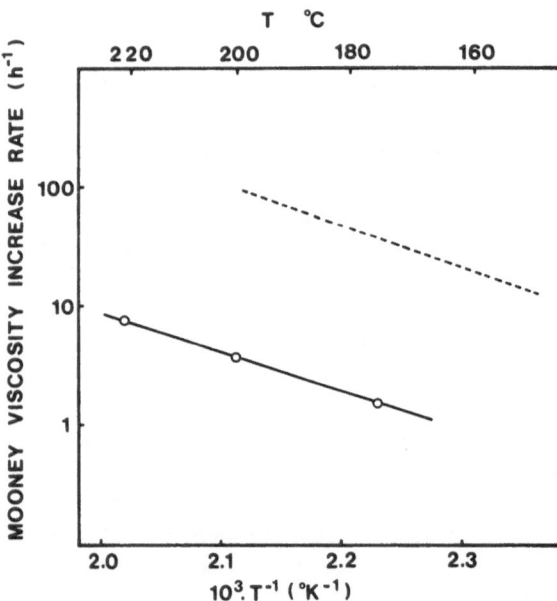

Fig. 28. Self-condensation Arrhenius plot of HTI and conjugated diene butyl rubber (Ref.[8]) (broken line)

(CDB)[8]. Furthermore, the value of E_{app} is in the range of data obtained for the activation energy pertaining to the Diels-Alder dimerizations of several non-cyclic conjugated dienes[83] and this agreement supports the existence of homocondensation Diels-Alder reactions dominating during gel formation.

Figure 28 shows, moreover, that the dienes unsaturation present in CDB[8] are more reactive than those of HTI and this difference is due to entropic factors originating, in turn, from the structural difference of the conjugated systems of HTI and CDB. In fact, the latter copolymer contains usually about 0.6 mole % of diene systems, very reactive in Diels-Alder reactions, having *trans-trans* and *exo-trans* configuration which are considered more reactive, for steric reasons, than other structures[8].

As EPTMs containing dienic unsaturations easily undergo Diels-Alder reactions, isobutene-triene copolymers are also subjected to viscosity increase on prolonged storage as a consequence of a slow but progressive self-curing occuring even at room temperature.

d 2) Reactions with Monofunctional Dienophiles. Strong dienophiles, such as maleic anhydride, react with HTI transforming its diene system completely within three hours at 120 °C, by working in solution and with a molar excess of dienophile. The reaction course is evident by the progressive disappearance of the UV absorption band at 235 mm, typical of conjugated unsaturation.

By contrast, OTI reacts incompletely, even at 150 °C. The block distribution of 2,4,6-octatriene in OTI seems the main reason for the incomplete reaction of the adjacent diene groups, because Bell observed[80] largely incomplete reactions of poly-1,3,5-hexatriene with several dienophiles; in particular, with bulky ones, more than 50% conversion was never achieved. In fact, severe steric constraints arise for a diene group belonging to a 1,6-triene homosequence when the two adjacent Diels-Alder adducts are formed.

The presence of carboxylic groups, introduced by maleic anhydride, allows modified HTI to cross-link with unconventional agents, e. g. polyvalent metal oxides, hydroxides, polyamines. For instance, a sample of HTI, completely reacted with maleic anhydride, has been cured with ZnO at 153 °C for 20 minutes and showed interesting tensile properties. Furthermore, the dynamic-mechanical spectra (Fig. 29)

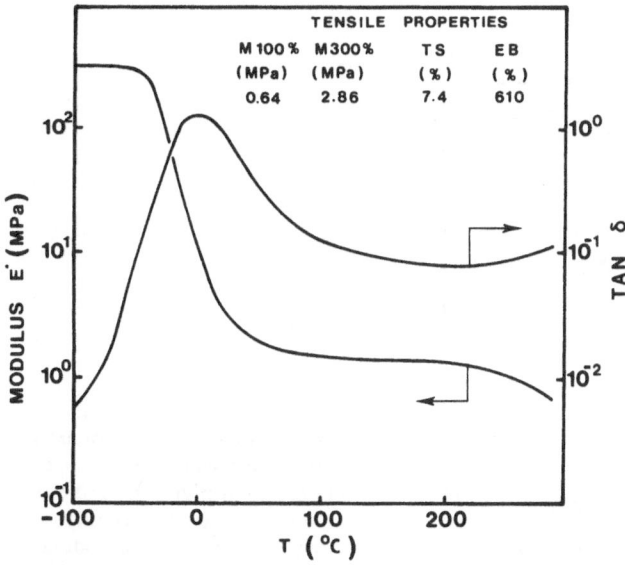

Fig. 29. Dynamic mechanical spectrum (at 110 Hz) of HTI (HT = 1.95 mol %) reacted with maleic anhydride and vulcanized with ZnO (5 %) at 153 °C for 20 min

at low temperatures produce a pattern very similar to that of butyl rubber[84], while between 50 and 200 °C a rubbery plateau exists. Beyond 200 °C ionic clusters begin to disintegrate and the polymer flows. The same upper temperature value was found for other carboxylate elastomers treated with ZnO[85].

d 3) Reactions with Bifunctional Dienophiles. Since the reaction of a polymer containing conjugated unsaturations with a bifunctional dienophile achieves an unconventional curing, the cross-linking process can be studied by means of a curometer and by measuring the torque moment of the polymer vs. time. The results obtained when HTI or OTI are compounded with a stoichiometric amount of p-phenylene-N,N′-bis-maleinimide (BMI) and heated at 160 °C, are plotted in Fig. 30. Once again the lower reactivity of OTI with respect to HTI, which originates from differences of structure between the two copolymers, is evident.

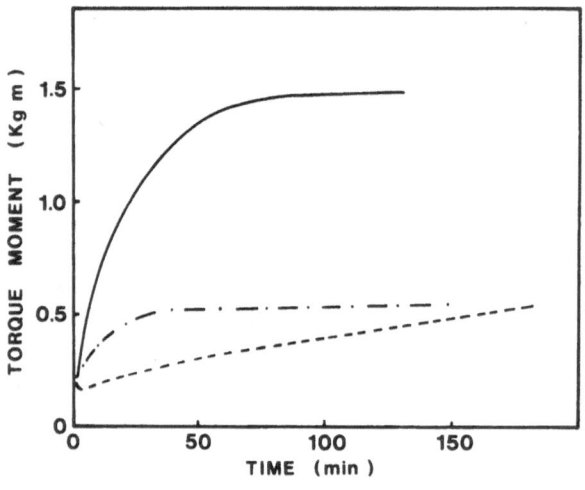

Fig. 30. Vulcanization kinetic curves of some isobutene-triene copolymers cured with BMI (equimolar amounts) at T = 160 °C; (—) = HTI (HT = 1.80 mol%); (– · –) = HTI (HT = 0.41 mol%); (---) = OTI (OT = 2.20 mol%)

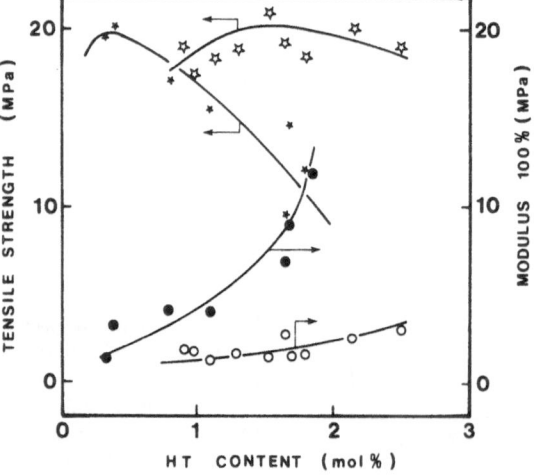

Fig. 31. Tensile properties vs. triene content of HTI vulcanized with sulfur recipe (○, ☆; recipe as in Fig. 23) or BMI (●★; conditions as in Fig. 30). (☆, ★) = Tensile strength; (○, ●) = 100% modulus

The properties of the vulcanizate (tensile properties vs. the content of unsaturation (Fig. 31)) indicate that a rigid network, based on C-C links, has been originated by the action of BMI. Such a network allows one to obtain tensile properties similar to those displayed by conventional sulfur vulcanizates, only when the content of unsaturation is low. On the other hand, BMI yields cross-linked HTI with high vulcanization kinetics which exhibits the same aging properties as butyl rubber cured with alkylphenol-formaldehyde resins[86]. The best set of properties are obtained with HTI containing less than 0.5 mol % of 1,3,5-hexatriene and using a molar ratio dienophile/unsaturation = 2 ÷ 3. The thermal-oxidative resistance of HTI vulcanized by BMI is noteworthy (Fig. 32) and accounted for by the formation of cyclohexenic rings inserted in the chains, so that the rupture of the cyclohexene double bond does not imply the rupture of the chain.

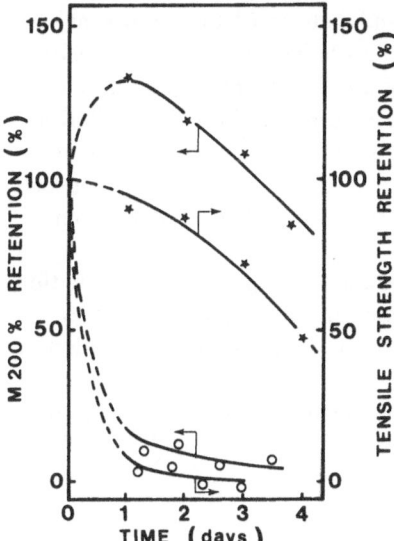

Fig. 32. Tensile properties retention vs. aging time of HTI (*) and IIR (o). Conditions: 170 °C in air

4. Chlorination and Oxidation Reactions

Polymer halogenation is of great practical interest because it permits the modification of the physical properties of the original polymer and also its chemical behavior, for instance during cross-linking processes.

Halogenation of saturated hydrocarbon polymers can hardly be controlled and is frequently associated with chain degradation phenomena[87]. In contrast, the presence of randomly distributed olefinic unsaturations, allows selective halogenation reactions by adopting appropriate conditions. For instance, butyl rubber can be chlorinated or brominated in allylic positions and chloro-butyl or bromo-butyl rubber results[7]. The latter polymers are very interesting since they exhibit fast curing rates when sulfur and ZnO are introduced in the formulations.

Copolymers containing conjugated double bonds are expected to react with halogens more easily and selectively than copolymers containing olefinic double bonds because of the high reactivity of dienic systems. Furthermore, the possibility of 1,4-addition of the halogen molecule to the diene system allows halogen atoms to enter the allylic positions and hence to create very reactive sites in the resulting polymer. Also the addition of halogenidric acids to a conjugated diene system should be feasible.

The results obtained by reacting HTI with solutions of Cl_2 or HCI are reported in Table 25 and confirm the expections. Chlorine reaction is very fast under mild conditions and within ten minutes the UV absorption band of the dienic system disappears. However, the amount of chlorine fixed by the polymer is slightly dependent on the quantity of the reagent employed. Furthermore, the intrinsic viscosity data indicate that the polymer degradation is almost negligible in $CHCl_3$ and slightly evident when the reaction is carried out in n-heptane.

Chlorination experiments, carried out with the aim of clarifying the structure of the resulting products, were performed on 2,4-hexadiene (HD), used as a model compound, by employing equimolar amounts of Cl_2 and diene in $CHCl_3$ solution at $-20\ ^\circ C$. MS-VPC analysis of the chlorinated products indicated that two derivatives (relative ratio 2.7:1) resulted from the addition of one Cl_2 molecule per dienic unit. Small amounts (less than 10%) of tri-chlorinated-HD were also formed. ^{1}H-NMR investigations showed the prevalent formation of 1,4-adducts (signals centered at 5.7, 4.4 and 1.5 ppm from HMDS attributable, respectively, to $-CH=$, $-CHCl-$ and CH_3 groups) and minor amounts of 1,2-adducts (weak signals at 1.7, 2.1 and 4.1 ppm assigned to CH_3-$C=$ and $-CHCl-$ groups).

The reaction of HT with HCl is relatively slow and incomplete at $-20\ ^\circ C$ within 1.5 hours, while a slight increase in the intrinsic viscosity of the polymer may indicate that some cross-linking reaction has occurred under the protonating action of HCl.

The insertion of the chlorine atoms into allylic positions is indirectly supported by the behavior of chlorinated HTI during vulcanization kinetics (Fig. 33) compared

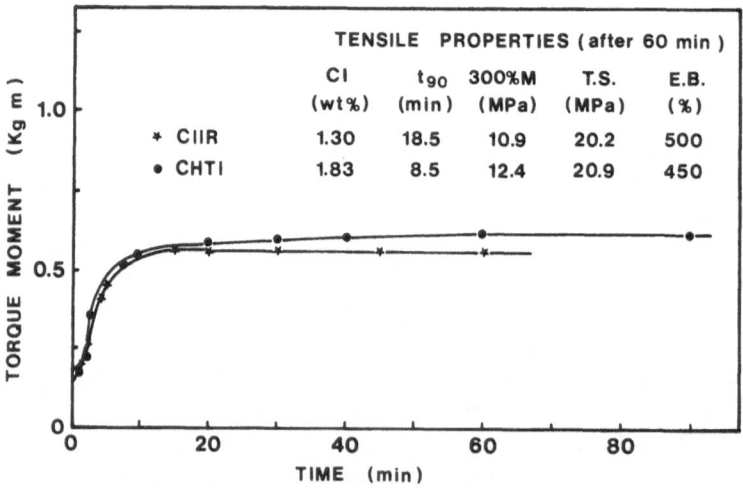

Fig. 33. Vulcanization kinetics and tensile properties of chlorinated HTI (CHTI, •) and comparison with chlorobutyl rubber (CIIR, *). Recipe as in Fig. 23 except, MBT = 1; T = 160 °C

Table 25. Chlorination of isobutene-1,3,5-hexatriene copolymer[a]

Solvent	Chlorination agent		T (°C)	Time (min)	Conv.[c] (%)	Chlorinated copolymer	
	(Type)	(Molar ratio)[b]				Cl content (%)	$[\eta]$ (dl/g)
n-Heptane	Cl_2[d]	1.3	25	10	97	1.83(1.47)[f]	2.44
CHCl3	Cl_2[d]	1.5	−20	10	98	2.23(1.47)[f]	2.52
CHCl3	HCl[e]	n.d.	−20	90	35	0.62(0.73)[f]	2.84

a 1,3,5-hexatriene content = 1.65 mol %; $[\eta]$ = 2.56 dl/g.
b Chlorination agent/hexatriene.
c Determined through the decrease of UV absorption at 235 nm.
d CCl4 solution.
e An excess of HCl was bubbled into the copolymer solution.
f Calculated value for a stoichiometric addition.

with that of chlorobutyl rubber. Also the tensile properties of the polymer, vulcanized with sulfur and ZnO, confirm that a network, similar to that of cross-linked chlorobutyl rubber has been formed.

Linear dienic conjugated systems were found very reactive also toward oxygen. Some model compounds, i. e. (III,a), 2,4-hexadiene and 2,5-dimethyl-2,4-hexadiene, dissolved in n-heptane, reacted quickly with a stream of oxygen at 75 °C (Fig. 34). Reference was made to an olefinic model compound, i. e. NB, which exhibited practically no reactivity under the adopted conditions. Figure 34 shows that the higher the methyl content of the model compound the faster the reaction with O_2.

By eliminating the volatile components under vacuum, the reaction mixture yields oleous products which have been subjected to elemental and physico-chemical analyses (Table 26). The viscous oil products show an oligomeric character and a structure originating from the alternating copolymerization of an O_2 molecule with the conjugated diene system. In fact, the absorption bands of the dienic systems are

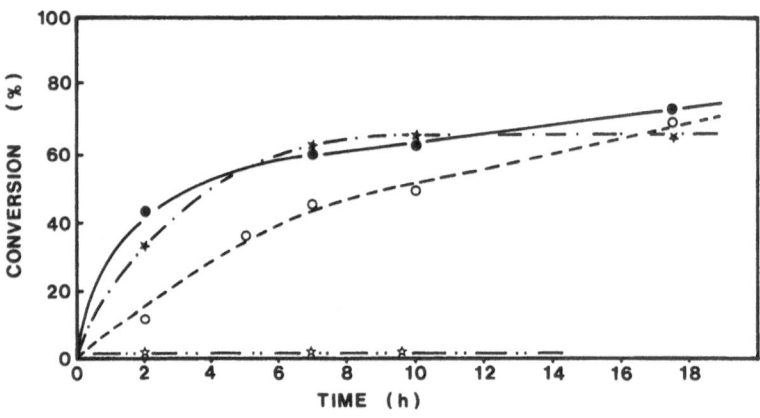

Fig. 34. Reactions with oxygen of some model compounds: conversion-time curves obtained by VPC analysis (III,a) = (•); HD = (○); DMHD = (⋆); NB = (☆)

Table 26. Analysis of the products resulting from the reaction of some conjugated dienes with oxygen

Type of diene	Elemental analysis C %	H %	Calculated structure	$\overline{M}n$	$\overline{DP}_n$ [c]
(III,a)	79.4	9.6	$(III,a)_1(O_2)_{0.68}$	460	2.2
HD[a]	64.3	9.3	$(HD)_1(O_2)_{0.92}$	820	7.2
DMHD[b]	69.2	10.2	$(DMHD)_1(O_2)_{0.90}$	540	3.8

[a] 2,4-Hexadiene.
[b] 2,5-Dimethyl-2,4-hexadiene.
[c] Calculated for an alternating copolymer.

practically absent in the UV spectra of all the oligomers of Table 30. ^{1}H-NMR analysis confirms a noticeable shift of all the signals with respect to the initial model compounds. The modifications of the ^{1}H-NMR spectra are interpretable for all the products of Table 26, on the basis of a main 1,4-opening of the diene system followed by the insertion of an O_2 molecule to yield the structure

$-\overset{|}{C}-\overset{|}{C}=C-\overset{|}{C}-O-O-$. The calculation of the signals intensity agrees with the pro-

posed enchainment.

Therefore, the behavior of the model compounds simulating the reactive sites existing on (III)-EPTM, HTI and OTI, predicts a significant reactivity of the corresponding dienic systems toward molecular oxygen when they are heated at moderate temperature in the absence of anti-oxidants and light. Very likely, free radicals, originating from peroxide or hydroperoxide decomposition, attack the conjugated unsaturations of the model compound molecules (1,4-addition mechanism) which are quenched by a molecule of oxygen. The less the dienic group is hindered the longer the lenght of the oligomeric chain. The formation of these substantially alternating copolymers has already been reported for some linear[90, 91] and cyclic[90] conjugated dienes. In particular, Kern[91] obtained polyperoxides from 2,3-dimethyl-butadiene with 1,2-enchainment of the dienic system, whereas Hock[92] found the 1,4-structure in polyperoxides obtained from cyclohexadiene.

IV Summary and Conclusions

The principles determining the synthesis of hydrocarbon copolymers containing a system of conjugated double bonds, randomly distributed, in general, along the chains, have been discussed. Two wide classes of polymers can be obtained under suitable conditions:
a) ethylene-propylene-based terpolymers (EPTM)
b) copolymers of isobutene with tri-conjugate linear trienes.

Coordination catalysts displaying low acidity are used for synthesizing the polymers of the former class, while cationic catalyst systems are necessary for the latter class.

The diene functionality introduced in the polymers described in this work can have a different structure, namely:
1) endocyclic conjugated double bonds (c. d. b) present as pendant groups, as in the case of (V) or (VI)-based EPTMs;
2) exocyclic c. d. b., present as pendant group in the chains, as observed in (I) or (III)-based EPTMs;
3) linear c. d. b., again present as pendant group, as in the case of (VII)-EPTM;
4) linear c. d. b. inserted in a polyisobutene chain and originating from the 1,6-opening of triconjugated trienes.

In all the cases, except group 4), diene systems carry substituting methyl groups which prevent the unsaturated system from reacting during the polymerization step thereby reducing side-reactions, in particular Diels-Alder condensations in the case of group 1 and 3.

The reactivity of the diene groups above listed during typical reactions, i. e. with free radicals, chlorinating or oxydizing agents, during Diels-Alder homocondensations or reactions with dienophiles, has been investigated with both polymer and low molecular weight model compounds in order to elucidate the reaction mechanism of the reactions considered.

The results obtained allow us to conclude that in several cases the polymers investigated possessed a reactivity which is qualitatively and quantitatively different from that shown by the corresponding polymers containing monoenic unsaturation. Furthermore, the peculiar reactivity of each dienic group allows the selection of a functionality more suitable for the post-modification reaction which is desired.

In particular, c. d. b. of group 1) when containing the cyclopentadienyl system show an enhanced reactivity in normal sulfur vulcanization processes. The analysis of the type of network formed has shown a prevalence of C–C and C–S–C links which impart high values to the tensile properties and improved resistance to high temperature. The possibility of covulcanizing with highly unsaturated elastomers, which is usually not achieved with EPDMs, comes from the high vulcanization kinetics of these EPTMs. Furthermore, terpolymers containing cyclopentadienylic unsaturations are very prone to undergo Diels-Alder reactions, in particular with classical dienophiles, which introduce polar pendant groups (–COOH, –CN, $>C = O$, etc.) in the hydrophobic chains.

A peculiar behavior is displayed by terpolymers containing c. d. b. of group 2) in the presence of free radicals. Highly efficient radical grafting or radical curing reactions are thus obtained. Investigations carried out with model compounds have shown that the reaction mechanism involves mainly allylic hydrogen abstraction (in the presence of phenyl and phenylcarboxy or t-butoxy radicals) and addition to the dienic system by cyanoisobutiric radicals. Usually a low amount of t-alkyl radicals is formed on the saturated backbone of the terpolymer, and the β scission reaction which follows t-alkyl radical formation accounts for some chain degradation which is observed during radical grafting.

Polymers containing c. d. b. belonging to groups 3) and 4) show chemical reactivity which is intermediate with respect to groups 1) and 2), since their dienic system can assume the cisoid conformation responsible for Diels-Alder reactions (however, in this respect they are less reactive than copolymers of group 1), while the presence of reactive allylic hydrogen atoms allows to easy radical grafting and curing reactions. In particular, isobutene copolymers containing unsaturations of group 4) can be cross-linked in the presence of peroxides, whereas isobutene-isoprene copolymers undergo relevant degradation processes under similar conditions.

Conjugated diene systems display noticeable reactivity in other reactions, e. g. toward halogens, halogen hydracids, or oxygen. Experiments with model compounds have elucidated the mechanism of these reactions which, in the case of chlorine and oxygen, take place mainly through 1,4-addition. Therefore, the resulting polymers contain very reactive chlorine atoms in allylic positions, suitable for curing processes. Moreover, cross-linking reactions are feasible in the presence of molecular oxygen on the basis of the behavior of the dienic model compounds.

In conclusion, the classes of polymers described in this work offer interesting possibilities for modifying hydrocarbon macromolecules through specific reactions involving the chemistry of conjugated diene systems, and new polymers and materials can be obtained for useful and specific applications.

Acknowledgments. We wish to thank Drs. S. Arrighetti, A. Brancacio, G. Ghetti, C. Locati, F. Mistrali and E. Vajna who collected several results reported in this work. Thanks are due also to Editrice di Chimica S.p.A. (Milan) for the permission to reproduce Figs. 3, 4 from Ref. 21,b and to John Wiley & Sons, Inc. (New York) for Figs. 15–17 and 20 taken from Ref. 9.

V References

1. Natta, G., Mazzanti, G., Corradini, P.: Atti Accad. Naz. Lincei, Red. Classe, Sci. Fis. Mat. Nat. *25* (1–2) 3 (1958)
2. Chauser, M. G.: Russ. Chem. Rev. *45*, 348 (1976)
3. Bell, V. L.: J. Polymer Sci. A2, 5291 (1964)
4. Baker, W. P.: J. Polymer Sci. A-1, 655 (1963)
5. Krentsel, B. A., Mushiva, E. A., Kharkova, E. M., Shishkina, M. V.: Eur. Polymer J. *11*, 865 (1975)
6. Stefanovskaya, N. N., Shmonina, V. L., Gavrilenko, I. F., Tinyakova, E. I., Dolgoplosk, B. A.: Dokl. Akad. Nauk SSSR *174*, 1356 (1967)
7. Baldwin, F. P., Buckley, D. J., Kuntz, I., Robinson, S. B.: Rubber Plastic Age *42*, 500 (1961)
8. Baldwin, F. P., Gardner, I. J., Malatesta, A., Rae, J. A.: Paper presented at 108th A. C. S. Rubber Div. Meeting, New Orleans, October 7–10, 1975; Rubber Chem. Technol. *49*, 390 (1976)
9. Cesca, S.: J. Polymer Sci., Macromol. Rev. *10*, 1 (1975)
10. Priola, A., Corno, C., Cesca, S.: in preparation
11. 1970 Book of ASTM Standards, Pt. 28, p. 607, Philadelphia, Pa.: American Society for Testing and Materials, 1970
12. Cesca, S., Arrighetti, S., Marconi, W.: Chim. Ind. (Milan) *50*, 171 (1968)
13. Ghetti, G., Corradini, G., Bulla, V., Bruzzone, M.: Chim. Ind. (Milan) *51*, 1361 (1969)
14. Cesca, S., Roggero, A., Salvatori, T., De Chirico, A., Santi, G.: Makromol. Chem. *133*, 161 (1970)
15. Cesca, S., Bertolini, G., Santi, G., Duranti, P. V.: J. Polymer Sci. A-1, *9*, 1575 (1971)
16. Cesca, S., Bertolini, G., Santi, G., Roggero, A.: J. Macromol. Sci. A–7, 475 (1973)
17. Cesca, S., Duranti, P. V., Roggero, A., Santi, G., Bruzzone, M.: J. Polymer Sci., Chem. Ed. *12*, 1209 (1974)
18. Cesca, S., Arrighetti, S., Bertolini, G., Bruzzone, M.: J. Macromol. Sci., Chem. A-*8*, 393 (1974)
19. Cesca, S., Arrighetti, S., Priola, A., Duranti, P. V., Bruzzone, M.: Makromol. Chem. *175*, 2539 (1974)
20. Cesca, S., Priola, A., Santi, G.: Polymer Letters *8*, 573 (1970)
21. Arrighetti, S., Brancaccio, G., Cesca, S., Giuliani, G. P.: Chim. Ind. (Milan) *59*, 483, 605, 685 (1977)
22. Musco, A., Silvani, A.: J. Organometal. Chem. *88*, C41 (1975)
23. Priola, A., Ferraris, G., Di Maina, M., Giusti, P.: Makromol. Chem. *176*, 2271 (1975)
24. Wassermann, A.: J. Chem. Soc. *1936*, 1028 and *1939*, 375
25. Atarot, H., Faucitano, A., Cesca, S.: Eur. Polymer J. *12*, 169 (1976)
26. Baldwin, F. P., Borzel, P., Makowski, H. S.: Rubber Chem. Technol. *42*, 1167 (1969)
27. Pearson, D. S., Böhm, G. G. A.: Rubber Chem. Technol. *45*, 1, 193 (1972)
28. Sheehan, C. J., Bisio, A. L.: Rubber Chem. Technol. *39*, 149 (1966)
29. Natta, G., Mazzanti, G., Valvassori, A., Sartori, G., Barbagallo, A.: J. Polymer Sci. *51*, 429 (1961)

30. Kennedy, J. P.: J. Polymer Sci. A–1 6, 3139 (1968)
31. Kennedy, J. P.: J. Macromol. Sci. Chem. A–6, 329 (1972)
32. Cesca, S., Bruzzone, M., Priola, A., Ferraris, G., Giusti, P.: Rubber Chem. Technol. 49, 937 (1976)
33. Tokita, N., Scott, R.: Rubber Chem. Technol. 46, 1019 (1973)
34. Gardner, I. J., Ver Strate, G.: Rubber Chem. Technol. 46, 1019 (1973)
35. Sonnenfeld, R. J., Kahle, G. R., Buck, O. G.: J. Appl. Polymer Sci. 13, 365 (1969)
36. Sauer, J.: Angew. Chem., int. ed. 6, 16 (1967)
37. Valvassori, A., Sartori, G., Mazzanti, G., Pajaro, G.: Makromol. Chem. 61, 46 (1963)
38. Gardner, I. J., Baldwin, F. P.: Paper presented at 108th A. C. S. Rubber Div. Meeting, New Orleans, October 7–10, 1975; Rubber Chem. Technol. 49, 391 (1976)
39. Ital. Pat. Appl. 26222, August 8, 1975, filed by Snamprogetti S. p. A.
40. Thaler, W. A., Buckley, D. J.: Rubber Chem. Technol. 49, 960 (1976)
41. Kennedy, J. P., Canter, N. H.: J. Polymer Sci. A–1, 5, 2455, 2712 (1967)
42. Ceresa, R. J.: Block and graft copolymers. London: Butterworth, 1962
43. Recent advances in polymer blends, grafts and blocks. Sperling, L. H. (ed.). New York: Plenum Press, 1974
44. Copolymers, polyblends and composites. Platzer, N. A. (ed.). Adv. Chem. Ser. 142, 141–231 (1975)
45. Placek, C.: Chemical Process Review n° 46, Park Ridge, New Jersey: Noyes Data Co., 1970
46. Meredith, G. L.: Rubber Chem. Technol. 44, 1130 (1971)
47. Allen, P. W., Merrett, F. M.: J. Polymer Sci. 22, 193 (1956)
48. Locatelli, J. L., Riess, G.: Angew. Makromol. Chem. 32, 117 (1973)
49. Bevington, J. C.: J. Chem. Soc. 1954, 3707
50. Rajbenbach, A., Szwarc, M.: J. Am. Chem. Soc. 79, 6343 (1957)
51. Rajbenbach, A., Szwarc, M.: Proc. Roy. Soc. (London) A 251, 1266 (1959)
52. Gresser, J., Rajbenbach, A., Szwarc, M.: J. Am. Chem. Soc. 82, 5820 (1960)
53. Pulman, B.: J. Chem. Phys. 55, 790 (1958)
54. Hefter, H. J., Hecht, I. A., Hammond, G. S.: J. Am. Chem. Soc. 94, 2793 (1972)
55. Faucitano, A., Faucitano Martinotti, F., Buttafava, A., Cesca, S.: Eur. Polymer J. 12, 421 (1976)
56. Faucitano, A., Buttafava, A., Faucitano Martinotti, F., Cesca, S.: J. Phys. Chem. 81, 354 (1977)
57. Van Drumpt, J. D., Oosterwijk, H. H.: J. Polymer Sci., Chem. Ed. 14, 1495 (1976)
58. Loan, L. D.: J. Polymer Sci. A–2, 2127 (1964)
59. Bruzzone, M.: Hydroc. Process. 52, 153 (1973)
60. Bateman, L.: The chemistry and physics of rubber-like substances, p. 486. London: Mc Laren and Sons, 1963
61. Pryor, W. A., Fuller, D. L., Stanley, J. P.: J. Am. Chem. Soc. 94, 1632 (1972)
62. Walling, C., Thaler, W.: J. Am. Chem. Soc. 83, 3877 (1973)
63. Baldwin, F. P., Borzel, P.: Rubber Chem. Technol. 43, 552 (1970)
64. Loan, L. D.: J. Polymer Sci. A–2, 3053 (1964)
65. Baldwin, F. P., Ver Strate, G.: Rubber Chem. Technol. 45, 709 (1972)
66. Bateman, L., Glazebrook, R. W., Moore, C. G., Porter, M., Ross, W., Saville, R. W.: J. Chem. Soc. 1958, 2838, 2846, 2866
67. Saville, D., Watson, A. A.: Rubber Chem. Technol. 40, 100 (1967)
68. Hoffmann, W.: Vulcanization and vulcanizing agents, p. 21. London: Mac Laren and Sons, 1967
69. Walker, J.: J. of IRI 8, 64 (1974)
70. Cotten, G. R.: Rubber Chem. Technol. 45, 129 (1972)
71. Block, G. A.: Organic accelereators in vulcanization of rubbers, p. 224. Jerusalem: I. P. S. I., 1968
72. Wassermann, A.: Diels-Alder reactions, p. 52. Amsterdam: Elsevier Publ. Co., 1965
73. Benford, G., Wassermann, A.: J. Chem. Soc. 1939, 362
74. Benford, G., Khambata, S., Kauffmann, H., Wassermann, A.: J. Chem. Soc. 1939, 381

75. Kaufmann, H., Wassermann, A.: J. Chem. Soc. *1939*, 870
76. Bouchal, K., Toupek, J., Pokorny, S., Hzabak, F.: Makromol. Chem. *137*, 95 (1970)
77. Huisgen, R., Grashey, R., Sauer, J.: The chemistry of alkenes, Patai, S., (ed.), Vol. I, p. 739. New York: Interscience, 1964
78. Wassermann, A.: Trans. Faraday Soc. *34*, 128 (1938)
79. Wassermann, A.: Ber. *66*, 1933 (1932)
80. Bell, V. L.: J. Polymer Sci. A—*2*, 5305 (1964)
81. Platé, N. A.: Pure and Appl. Chem. *46*, 49 (1976)
82. Langley, R. N.: Macromolecules *1*, 348 (1968)
83. Onishenko, A. S.: Diene synthesis, p. 61. Jerusalem: I. P. S. R., 1964
84. Ferry, J. D.: Viscoelastic properties of polymers, Chapt. 12. New York: Interscience, 1970
85. Tobolsky, A. V., Lyons, P. F., Hata, N.: Macromolecules *1*, 515 (1968)
86. Hoffmann, W.: Vulcanization and vulcanizing agents, p. 300. London: Mac Laren and Sons, 1967
87. Bruzzone, M., Crespi, G.: Chim. Ind. (Milan) *42*, 1226 (1960)
88. Flory, P. J.: Principles of polymer chemistry, p. 358. Ithaca, New York: Cornell University Press, 1953
89. Brydon, A., Burnett, G. M., Cameron, G. G.: J. Polymer Sci. Chem. Ed. *11*, 3255 (1973)
90. Bodendorf, K.: Arch. Pharm. *271*, 1 (1933)
91. Kern, W., Stolman, I.: Makromol. Chem. *7*, 199 (1951)
92. Hock, H., Depka, F.: Ber. *84*, 349 (1951)
93. British, P. 1.020.670, assigned to SNAMPROGETTI S.p.A.
94. U. S. P. 3.681.309, assigned to ESSO Co.
95. Faucitano, A., Atarot, H., Faucitano Martinotti, F., Comincioli, V., Arrighetti, S., Cesca, S.: submitted to Eur. Polymer J.
96. Gear, C. K.: Numerical initial value problems in ordinary differential equations, New York, Englewood Cliffs: Prentice Hall, 1971
97. Faucitano, A., Atarot, H., Faucitano Martinotti, F., Comincioli, V., Cesca, S.: Eur. Polymer J., in press

Received December 15, 1978
G. Dall'Asta (editor)

Recent Advances in Membrane Science and Technology

V. T. Stannett and W. J. Koros

Department of Chemical Engineering North Carolina State University, Raleigh, North Carolina 27650, U.S.A.

D. R. Paul

Department of Chemical Engineering University of Texas, Austin, Texas 78712, U.S.A.

H. K. Lonsdale and R. W. Baker

Bend Research, Inc., Bend, Oregon 97701, U.S.A.

In this review dealing with recent advances in membrane science, the term "membrane" will be used to indicate any medium which acts as a barrier to transport into or out of a region, provides selective transfer of one species over another or regulates the transport of a material to its environment at a controlled rate. In addition to the common usage of the word "membrane" to indicate a dense polymer film, the above definition includes a variety of interesting cases such as highly porous ultrafiltration membranes and hydrophobic liquid membranes with selectivity properties which can be tailored by incorporation of materials which selectively complex with one of the species to be processed. The important topics of controlled release of chemicals from polymeric devices and removal of volatile monomers from addition polymers such as poly(vinyl chloride) and poly(acrylonitrile) are also treated here.

This review does not aim for an encyclopedic coverage of the rapidly expanding field of membrane science. The topics treated do, however, present a reasonable sampling of the current literature on membrane science. The subjects treated reflect directly the interests of the authors and many of the topics are fundamentally related to the fact that the membrane is a high polymer. Indeed, the transport and sorption theories presented provide parameters whose physical interpretation often is a useful complement to other methods of physical analysis of the polymeric state. The study of gas sorption and transport in rubbery polymers is perhaps the simplest example of such a case and will be treated first.

Table of Contents

Gas Transport and Sorption

Rubbers

A variety of approaches have been used over the years to describe the way in which low molecular weight species penetrate nonporous rubber membranes. Common to most of these theories, however, are the elements first suggested by Graham[1] in 1866. He postulated that the mechanism of the permeation process entailed solution of the gas in the upstream surface of the membrane, diffusion through the membrane by "colloidal diffusion" (likened to diffusion in liquids) followed by evaporation from the downstream membrane surface. The actual diffusion process involves a series of thermally activated jumps with a directional bias imposed by the presence of a negative gradient in concentration. A diffusional jump can be considered to consist of the localization of sufficient thermal energy in the polymer near the penetrant to open a momentary volume fluctuation of sufficient size to permit passage of the penetrant to a new position in the polymer[2]. The statistical weighing of these jumps, caused by the concentration gradient, is responsible for the directionality of transport.

In 1879, von Wroblewski[3] showed that the solution of gases in rubber followed Henry's law, Eq. (1), and he defined an absorption coefficient as the number of cubic centimeters of gas measured at STP which dissolved in 1 cubic centimeter of rubber at one atmosphere pressure. Combining this with Fick's law, Eq. (2) he showed that the steady state permeation flux through a membrane of thickness ℓ is given by Eq. (3):

$$C = k_D p \tag{1}$$

$$J = -D\frac{\partial C}{\partial x} \tag{2}$$

$$J = \frac{k_D D \Delta p}{\ell} = \frac{P \Delta p}{\ell} \tag{3}$$

$$P = k_D D \tag{4}$$

where C is the penetrant concentration in the polymer, D is the diffusion coefficient, k_D is the absorption coefficient and Δp is the pressure difference between the top and bottom faces of the membrane. The diffusion coefficient, D, was assumed to be independent of the concentration of gas in the membrane. P is the steady state permeability, and is effectively defined by Eqs. (3) and (4). These are still the basic sorption and transport equations used today for gases in rubbery polymers.

The diffusion coefficient for gases in rubbers has an Arrhenius form[4]

$$D = D° \exp[-E^*/RT] \tag{5}$$

The activation energy, E^*, can be interpreted as the previously mentioned thermal energy that must be concentrated in the polymer adjacent to a penetrant to open a passage of sufficient volume to allow the penetrant to execute a diffusional jump.

An excellent review of the various theoretical expressions of this physical fact has been offered by Kumins and Kwei[5]. At a fixed temperature, many investigators have found that gas diffusion coefficients in rubbers can be correlated with the Lennard-Jones gas collision diameter. Michaels and Bixler[6] have shown, however, that even in the rubbery state, some orientation of anisotropic molecules such as CO_2 may occur during a diffusion step.

Early fundamental studies of gas transport in polymers were almost entirely confined to hydrocarbon materials above their glass transition temperatures. The essentially nonpolar structures of the elastomers led to a number of reasonably successful attempts to correlate gas transport parameters with various physical characteristics of the gases and the polymers. These have been summarized and discussed in a number of papers[7-9]. In addition to studies with hydrocarbon elastomers a few studies of other amorphous polymers above their glass transition temperatures have dealt with polyvinyl acetate[2, 10] silicones[11, 12] and fluorocarbon polymers[12, 13]. Recent studies have also dealt with poly(methyl acrylate)[14] poly-(vinyl methyl ether)[9, 15] and poly(vinyl methyl ketone)[9, 15]. With these more polar polymers, the correlations between the diffusivities and the physical properties of the polymers traditionally invoked in hydrocarbon elastomers tend to break down, and no clear correlations have been discovered for D or E* between the glass transition temperatures, the densities or the cohesive energy densities (CED) of these polymers. This is illustrated, for example in Table 1.

A major breakthrough in the study of gas and vapor transport in polymer membranes was achieved by Daynes in 1920[16]. He pointed out that steady-state permeability measurements could only lead to the determination of the product Dk_D and not their separate values. He showed that, under boundary conditions which were easy to achieve experimentally, D is related to the time required to achieve steady state permeation through an initially degassed membrane. The so-called "diffusion time lag", θ, is obtained by back-extrapolation to the time axis of the pseudo-steady-state portion of the pressure buildup in a low pressure downstream receiving volume for a transient permeation experiment. As shown in Eq. (6), the time lag is quantitatively related to the diffusion coefficient and the membrane thickness, ℓ, for the simple case where both k_D and D are constants.

Table 1. Relationship between the diffusion coefficient parameters for argon and the physical properties of four closely related polar polymers

Polymer	$D_{30\,°C}$ $\times 10^8$ cm^2/sec	E* Kcals/ mole	T_g °C	Density g/cc	C.E.D. cals/cc	$(\alpha \times 10^4)^a$ cc/g °C
Poly(vinyl methyl ketone)	0.32	18.4	18	1.19	127.5	4.6
Poly(vinyl acetate)	1.58	16.5	28	1.17	86.9	6.0
Poly(methyl acrylate)	5.7	14.8	3	1.22	94.5	5.6
Poly(vinyl methyl ether)	60.6	14.85	−23	1.05	81.3	4.7

a   Coefficient of cubical expansion.

$$D = \frac{\ell^2}{6\,\theta} \tag{6}$$

Since the product Dk_D is known from the steady state rate of permeation, k_D can also be obtained. This time lag method is the basis of most of the gas and some of the vapor transport studies made today. Little application of the time lag method was made until Barrer introduced the use of vacuum on the downstream side of the membrane and measured the gas permeation rate by monitoring the increase in pressure in a fixed downstream receiving volume[4]. Recently the original isobaric method has been reintroduced in a number of commercial permeability instruments.

Van Amerongen[17] showed experimentally that the absorption coefficients (now almost always termed the "solubility" or Henry's law constant) measured directly for various gases in rubber were the same as those obtained by taking the quotient of the permeabilities and the time lag diffusivities calculated from Eq. (6). Later, Frisch[18, 19] extended the analysis of the time lag to cases where the diffusion coefficients were concentration dependent and Eq. (6) no longer applies. Frisch[20] and Petropoulos[21] have also considered cases in which the diffusion coefficient is position or time dependent: in such cases the expressions relating the observed time lag to the transport parameters of the polymer are often very complicated.

The Henry's law constant, k_D, appearing in Eq. (1) characterizes the solubility of a gas in a rubber and is therefore related somewhat to the infinite dilution sorbate/

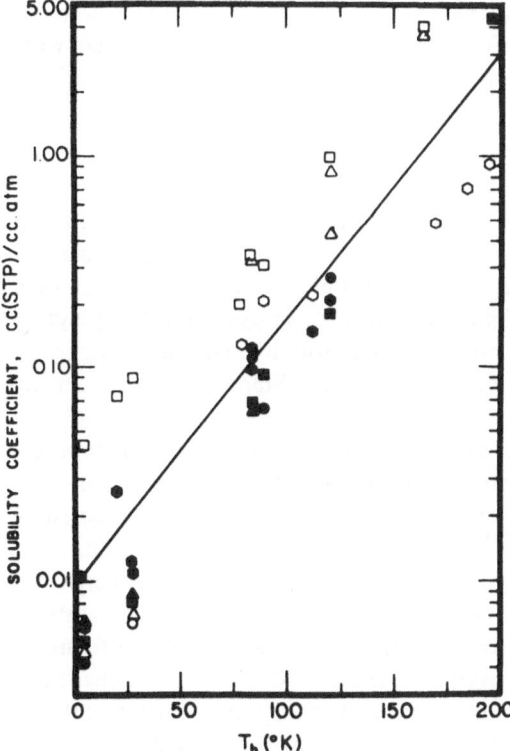

Fig. 1. Solubility coefficients for a number of gases in various polymers above their glass transition temperature. ● poly(methyl vinyl ether); ◕ poly(vinyl acetate); ▲ poly(methyl vinyl ketone); ■ poly-(methyl acrylate); ○ FEP; □ poly-(dimethyl siloxane); △ poly(phenyl siloxane). The solid line is that for natural rubber (Van Amerongen and Barrer, see Ref.[8], p. 65)

sorbent interaction. For a given rubber, useful correlations have been developed to
relate k_D to some parameter such as the normal boiling point of a gas which provides
a measure of the ease of condensing the gas. This correlation between the logarithm
of the solubilities and the boiling points of the gases also holds reasonably well for
polar polymers above T_g. All of the available data is presented in Fig. 1 together with
the original line of van Amerongen[7] and Barrer and Skirrow[22]. Stannett[8, 9] and
Bixler and Sweeting[23] have presented reviews of the various other schemes for correlating the Henry's law constants with these and other properties of the gases such
as the Lennard Jones potential well depth[6].

Stern and his co-workers[24] were among the first to systematically study the
transport and sorption of gases and vapors in high polymers at elevated pressures.
The data led them to apply the principle of corresponding states and to correlate
the logarithm of solubilities of a very large number of gases and vapors in polyethylene with the ratio of the critical temperature and the temperature of measurement. A further examination of the data spanning five orders of magnitude of solubilities showed an excellent linear relationship with $(T_c/T)^2$. The correlation included
the data of many other workers in addition to their own. A similarly close correlation
was later found for a large number of vapors in silicone rubber[25] and more recently
Stiel and Harnish applied this form of correlation to a large number of gases and
vapors in molten polystyrene[26]. The slopes and intercepts were somewhat different
in each of the above cases: the intercepts were -5.64, -5.10 and -1.02 and the
slopes were $+1.14$, $+1.08$ and $+1.18$ for polyethylene, silicone rubber and molten
polystyrene respectively. Approximately linear relationships have been found between the boiling point and the critical temperature and the Lennard-Jones potential
well depth[27]. It is not surprising, therefore, that the solubilities can be related, with
varying degrees of success, to all three of these parameters.

Glassy Polymers and Dual Mode Sorption and Transport

Fundamental studies of gas transport in polymers other than rubbers began with the
classical work of Meares in 1954[2]. He was the first to demonstrate and theorize
about the now well-known inflection in the Arrhenius plots of D near the glass transition temperature. He also speculated about two modes of sorption in glassy polymers.
Later studies were initiated with many polymers by Barrer, Michaels and their co-
workers together with important contributions by Brandt, Stern, Stannett and many
others[9, 28].

The concept of two or more modes of sorption of penetrants in polymers is very
familiar to cellulose and protein chemists for the case of water vapor. In fact combined Langmuir and Henry's law sorption was proposed and correctly formulated by
Matthes in 1944 for water in cellulose[29]. The discovery of dual mode sorption of
gases in glassy polymers and the subsequent realization that diffusion constants
determined by the time lag method did not have the same simple fundamental significance associated with these parameters for rubbery polymers was of profound
importance. Not only were the many carefully determined diffusion coefficients in
the literature of questionable value for polymers below their glass transition but a
good deal of the careful speculation about solution and diffusion and the effect of

the glass transition on gas transport in general was brought into question. The dual mode sorption of gases in glassy polymers was apparently first discussed by Meares[2, 10] in 1954 and studied in more detail in 1958. He measured the sorption of a number of gases in poly(vinyl acetate) both directly and from the quotient of the permeability and diffusion coefficients above and below the glass transition temperature. He postulated that glassy polymers contain more than the equilibrium amount of volume due to the presence of lower density regions constituting high energy sorption sites. The gases sorb in these "holes" exothermically. Also in 1958, Barrer, Barrie and Slater[30] independently arrived at somewhat similar conclusions for the sorption of C_4 and C_5 hydrocarbons in ethyl cellulose in the glassy state. They found the sorption isotherms concave to the pressure axis and formulated them as a combination of simple solution (Henry's law) and a Langmuir isotherm. The heats of mixing were found to be negative whereas the values calculated by Hildebrand's equations were always positive. Sorption in preexisting cavities was postulated to explain these differences. It is interesting that the sorption and diffusion data of Barrer, Barrie and Slater have been reanalyzed recently by Chan, Paul and Koros and found to closely conform to their modern treatment of dual mode sorption and transport[31].

The concept of dual mode sorption was first clearly demonstrated and quantified by Michaels, Vieth and Barrie in 1963[32]. The same authors also discussed its effect on the diffusion process itself. Vieth and his co-workers subsequently extended these findings to a number of polymer-gas systems and elaborated the theoretical aspects of the problem[33, 34]. In particular, a model for diffusion in glassy polymers, which has come to be known as the "total immobilization" model, was developed by Vieth and Sladek[33].

The basic assumptions of the dual mode sorption theory as it applies to the transport model of Vieth and Sladek, have been stated by Vieth et al. in their excellent review of the subject[34]. The sorption isotherm was described by the combination of a Henry's law "dissolved" component, C_D, and a Langmuir "hole filling" term, C_H, i.e.,

$$C = C_D + C_H = k_D p + \frac{C'_H b p}{1 + bp} \tag{7}$$

where k_D is the Henry's Law constant, p is the pressure and C'_H and b are the Langmuir capacity constant and affinity constant respectively. The transport model assuming complete immobilization has recently been considerably modified and this approach will be presented in some detail here.

The predicted effect of dual mode sorption on the time lag and permeability was derived by Paul[35] using the total immobilization transport model and experimentally verified by Paul and Kemp[36] using molecular sieves embedded in a silicone rubber. This was an excellent model system which fulfilled the postulate of complete immobilization of the Langmuirian mode penetrant. The possibility that gas molecules sorbed in the Langmuirian mode may not necessarily be completely immobilized in glassy polymers was first raised by Petropoulos in 1970[37]. Equations were developed and the possibility of these being used to check the assumption of immobilization by sorption and permeation data were described. The relaxation of the postulate of complete immobilization was highly significant in that it implied that true diffusion constants of dissolved gas could not be obtained in glassy polymers even with a combination of low pressure permeation and high pressure sorption equilibria data. The work of Petropoulos was extended by Paul and Koros in a paper published in 1976[38]. Expressions for the permeability and time lag were developed for varying degrees of immobilization ranging from complete to zero immobilization. Both the approach of Petropoulos[37] using chemical potential gradients and an approach using concentration gradients were developed and the differences compared. This model has come to be known as the dual mobility or partial immobilization model. The concentration gradient form of the transport expression is given in Eq. (8).

$$J = -D_D \frac{\partial C_D}{\partial x} - D_H \frac{\partial C_H}{\partial x} \qquad (8)$$

where D_D and D_H are the respective constant diffusion coefficients of the two sorbed populations C_D and C_H described in Eq. (7). Equation (9) which phenomenologically defines an effective concentration dependent diffusion coefficient in terms of the gradient in total concentration, C, is mathematically equivalent to Eq. (8):

$$J = -D_{eff}(C) \frac{\partial C}{\partial x} \qquad (9)$$

The difference between the two approaches is that Eq. (8) explicitly takes account of the unique character of gas sorption in glassy polymers and yields parameters with the potential for physical interpretation. The general theoretical prediction of the permeability for the partial immobilization model is given by Eq. (10):

$$P = k_D D_D \left[1 + \frac{FK}{1 + bp_2} \right] \qquad (10)$$

where $K = C'_H b/k_D$, $F = D_H/D_D$ and p_2 is the upstream driving pressure[38]. The downstream pressure is essentially equal to zero in most measurement situations and does not enter into Eq. (10). A detailed discussion of the significance of the various sorption parameters has been presented by Koros et al.[39]. As noted above, the parameter F equals the ratio of the diffusion coefficient of the gas in the Langmuir population, D_H, to the diffusion coefficient of the gas in the Henry's law population, D_D, i.e., $F \equiv D_H/D_D$. In the limit of F = 1, the permeability depends strongly on the upstream pressure, while in the limit of F = 0, the permeability is independent of the upstream pressure and is given by Eq. (4)[38]. A least squares fit of steady state permeability data versus the dimensionless variable $\frac{1}{1 + bp_2}$ yields D_D and F from the intercept and slope of Eq. (10). The parameters, b, k_D and K for use in Eq. (10) can be determined independently by a least squares fit of sorption data to the sorption isotherm in Eq. (7).

The combined procedure described above, which uses only sorption and steady state permeation data, specifies all five of the sorption and transport model parameters without requiring reference to the independently measured time lags. Comparison of theoretically predicted time lags with the experimentally measured values provides a rigorous test of the internal consistency of the transport and sorption data as well as a check of the applicability of the partial immobilization model for description of the transient processes.

The concentration dependent diffusion coefficient defined by Eq. (9) can be evaluated by differentiation of steady state permeation data without reference to the partial immobilization model[40]. The concentration dependent diffusion coefficient calculated from the partial immobilization model agrees very well with values calculated in this way, and one can consider them to be essentially identical mathematically[40]. The partial immobilization theory, therefore, serves to explain the source of the concentration dependency of D_{eff} in Eq. (9).

The partial immobilization prediction of the concentration dependency of D_{eff} is:

$$D_{eff} = D_D \left[\frac{1 + \dfrac{FK}{(1 + \alpha C_D)^2}}{1 + \dfrac{K}{(1 + \alpha C_D)^2}} \right] \qquad (11)$$

where the parameters $K = C'_H b/k_D$ and $\alpha = b/k_D$ are known from independent sorption experiments and C_D corresponds to the concentration of molecularly dissolved penetrant at the point of interest. Only in the case where there is no difference in

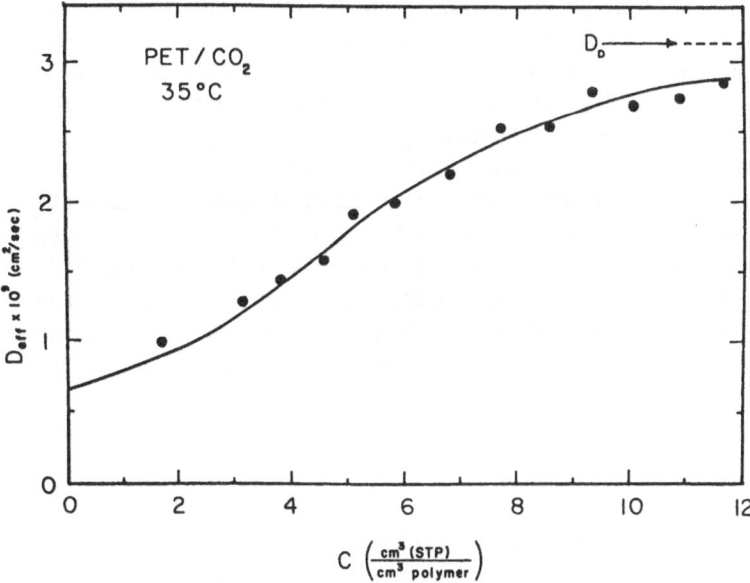

Fig. 2. Effective concentration dependent diffusion coefficient defined by Eq. (9). Note that D_{eff} asymptotically approaches D_D as C increases

inherent mobility of the two populations, i.e. $F = D_H/D_D = 1.0$, is D_{eff} independent of concentration. This simple case does not apply to carbon dioxide in PET as Fig. (2) shows clearly[41]. In fact, very significant deviations from equal mobility of the two species exist since the value of F ranges from 0.05 to 0.10 between 25 °C and 75 °C. The large difference in relative mobilities between gas in the two sorption modes was the basis for the earlier, so-called "total immobilization" transport model which assumes zero mobility for the Langmuir species, i.e. $D_H = 0$. This case, however, fails to account for any pressure dependence of the permeability and yields predicted time lags that are considerably larger than experimental values[42].

An analysis of gas transport for high upstream pressures in glassy polycarbonate at 35 °C suggests that anisotropic molecules such as CO_2 can sometimes execute primarily lengthwise oriented diffusion jumps in the glassy state, thereby greatly reducing their effective cross section[39]. Although the Lennard-Jones collision diameter is a widely accepted correlating parameter for diffusivities in relatively high mobility rubber or liquid media, a correlation based on this parameter was not successful in glassy polycarbonate because CO_2 appeared to have a much higher mobility than expected based on its collision diameter. It was found that the minimum effective cross-sectional diameter of a gas, as measured by the minimum diameter zeolite window that allows a particular gas to pass into the inner sorption cavity is a sensitive measure of the ability of a gas molecule to move in highly restrictive environments. For the gases, CO_2, CH_4, N_2, Ar, He, carbon dioxide is second only to helium in its ability to move through restricted environments[39]. This is probably related to the linearity of the CO_2 molecule; although a similar effect is not apparent for N_2.

A completely independent and parallel study made by Assink[43] using pulsed NMR for ammonia sorbed in polystyrene also suggests that partial immobilization occurs in this glassy polymer/penetrant system. The mobility of the Langmuirian population in this case was less than 5% as large as the mobility of the Henry's law mode. The data of Paul and Koros indicated mobilities of penetrant in the Langmuir mode relative to the Henry's law mode ranging from 8% for CO_2 up to 100% for He in polycarbonate[39] and from 6% to 10% for CO_2 in PET between 35 °C and 75 °C. At least in the systems studied, errors of less than a factor of two are found between the total and partial immobilization cases. In contrast the diffusion constants obtained by the simple time lag equation [Eq. (6)] are as small as one eighth of those estimated for the Henry's law mode diffusivity using the completly immobilized model (F = 0) in some cases. This means that actual errors of up to four fold in estimation of the true Henry's law mode diffusivity, D_D, can possibly be introduced by use of the simple time lag technique for glassy polymers if one fails to account for the effect of immobilization.

Gas Transport and the Glass Transition Temperature

The first measurements of gas transport in a polymer above and below the glass transition temperature, T_g, were those of Meares[2, 10] with poly(vinyl acetate). A clear break in the Arrhenius plots was found for all the gases investigated. A somewhat flatter part of the curve was found in the vicinity of the glass transition for the larger gas molecules. Similar measurements with oxygen were later performed by Stannett and Williams[44] and only a clean break at T_g was observed. Meares explained the observed transition region in terms of varying degrees of relaxation being involved in this region. The activation energies were always higher in the rubbery state than in the glassy state and differences of between one and five kilocalories per mole were observed with the larger penetrants exhibiting the greatest differences. Similar behavior has since been reported with gases in polyethylene terephthalate[32] poly-(methyl acrylate)[14] poly(methyl vinyl ketone)[15] poly(vinyl fluoride)[13] and poly-(vinylidene fluoride)[13]. With a vinyl chloride-acetate copolymer[45] and with poly-(vinyl chloride)[46] the break in the Arrhenius plot was only found for the larger penetrants.

Careful studies with a large number of gases in poly(ethyl methacrylate) by Stannett and Williams[44] showed no break in the Arrhenius plots of either P or D, even for large penetrants. However a later study of hydrogen and deuterium transport by Ziegl and Eirich[47], with the same polymer showed a clear break in the Arrhenius plots for both permeabilities and diffusivities. In view of this discrepancy in such an apparently unique system, the hydrogen transport in a sample which had been carefully annealed for two days at 10 °C above T_g, was remeasured[9] and did

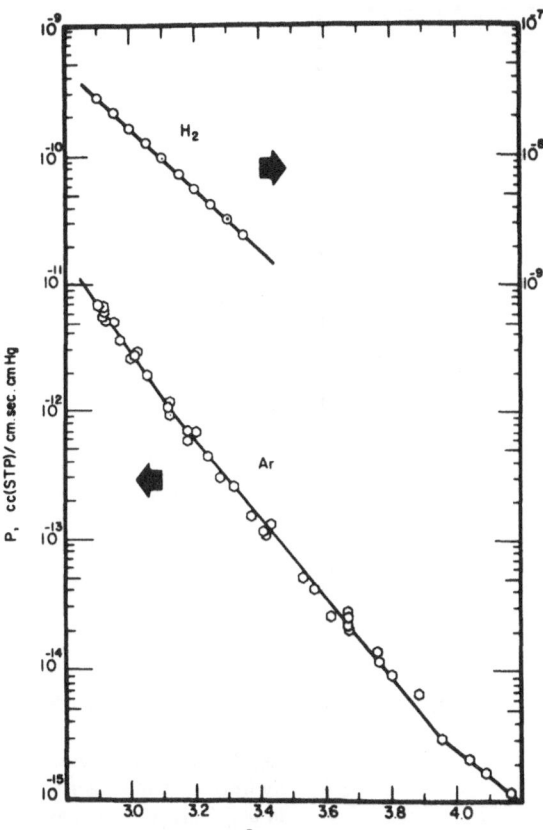

Fig. 3. The permeability of poly-(vinyl fluoride) to hydrogen[48] and argon[13] as a function of temperature

indeed show a small change in slope. The degree of change in slope at T_g is apparently related to the differences in the coefficients of thermal expansion above and below the glass transition[44].

Gas transport in poly(vinyl fluoride) also exhibits unusual behavior. Ziegl et al.[48] have measured hydrogen and deuterium transport in this polymer. Stannett and his colleagues have recently measured argon transport in the same polymer[13]. The results are presented in Figs. 3 and 4. The study by Ziegl et al.[48] shows no inflection at the glass transition temperature with the permeability plots but a sharp change of slope in the case of the diffusion data for both deuterium and hydrogen. The activation energies in the glassy state are however about 3 kcals per mole higher than in the rubbery state. This is the first and only example of such behavior which has been reported. The data of Stannett et al.[13] for argon however show the normal behavior with higher values of E^* in the rubbery state. It is interesting to note that the argon data also show a second transition at -15 °C. This is the first example of two transitions being revealed by gas transport and is in accordance with the dilatometry data of Enns and Simha[49] and the predictions of Boyer[50, 51]. The upper transition, 38 °C for argon and 45 °C for hydrogen and deuterium is believed by Enns and Simha[49] and by Boyer[50, 51] to be due to relaxation in amorphous regions

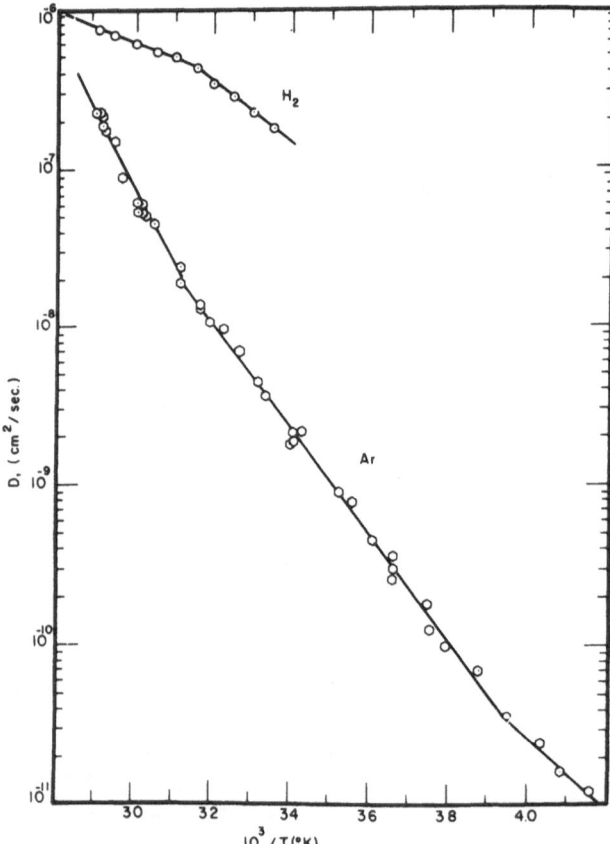

Fig. 4. The diffusivity of hydrogen[48] and argon[13] in poly(vinyl fluoride) as a function of temperature

restricted by the presence of the crystallites while the lower transition is that of the unrestricted amorphous part. These have been labeled Tg(u) and Tg(l) by those authors.

The above discussions have dealt with data obtained without the refinements introduced by the concept of partial immobilization. The low pressure limit of Eq. (11) corresponds to the effective diffusivity measured by low pressure time lag measurements, i.e.

$$\lim_{p_2 \to 0} \left[\frac{\ell^2}{6\theta} \right] = \lim_{p_2 \to 0} [D_{eff}] = D_D \frac{[1 + FK]}{[1 + K]} \tag{12}$$

Values calculated using this expression are plotted in Fig. 5 for the system CO_2/PET for comparison with the true Henry's law mode diffusion coefficient determined from a partial immobilization analysis[41]. D_D is, of course the only applicable diffusion coefficient above the glass transition temperature where Eq.(1) and (2) describe sorption and transport. The significant fact is that the change in slope of the Arrhenius plot of the apparent diffusion coefficient measured by time lags is actually *less* extreme than the break in slope of the plot of D_D which is related to the true change in mobility in the "normal" bulk polymer matrix upon traversing T_g[41]. The reasons for the lower activation energy for diffusion in the glassy state is not entirely clear. Presumably smaller segment sizes are involved in the diffusional process. This appears to be quite reasonable. It is interesting and relevant that the coefficient of thermal expansion of polymer is also less in the glassy state.

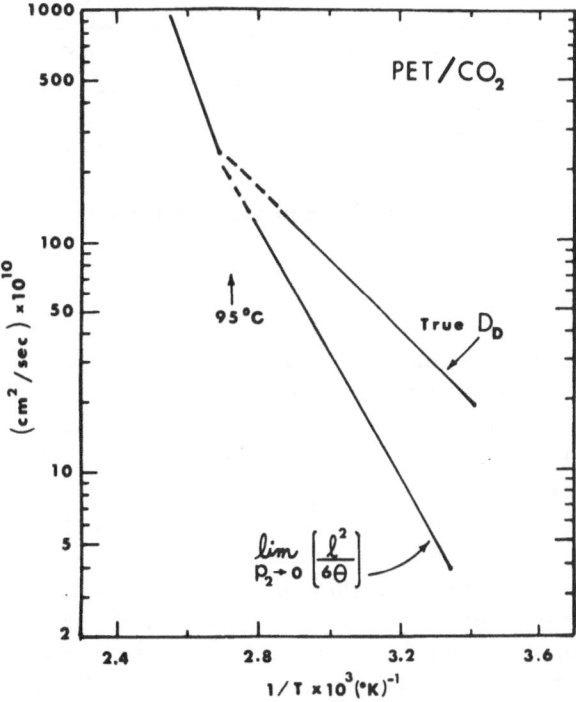

Fig. 5. Comparison of the true Henry's law mode diffusion coefficient, D_D, with the diffusion coefficient determined from time lag measurements at low pressure

Water Transport and Sorption

The basic principles developed for gases described in the previous sections also apply
to water and organic materials but the situation can be considerably more complicated.

Basic studies with water transport lagged behind those with gases for a number
of reasons. The measurements of the diffusion coefficients were quite difficult[52],
because the time lag method, almost universally used with gases, was not appropriate
with water due to concurrent adsorption on the walls of the receiving volume. The
quartz spring sorption balance method was also complicated in many cases by heat
generation or loss which accompanies water sorption or desorption in the case of
more highly sorbing polymers. These problems have been largely overcome by modifications to the equipment and by applying correction factors[52, 53]. It should be
noted that, on the other hand, the permeability coefficients are comparatively simple
to measure as are the equilibrium solubilities.

The fundamental behavior of water, both in transport and in sorption, is more
complex than with gases[54]. There is much more interaction between the penetrant
and the polymer leading, to permeability and diffusion coefficients which increase
with vapor pressure and concentration in cases where plasticization is an important
factor. In addition the high cohesive energy of water leads to the phenomenon of
clustering of water molecules in the polymer causing further complex behavior. Both
plasticization and clustering cause curving water vapor isotherms, convex to the pressure axis. In the case of glassy polymers a Langmuir component and a clustering or
Flory-Huggins type isotherm can be found[55, 56]. When strong hydrogen bonding
sites are present in the polymer a Langmuir component to the isotherm is also found.
With some polymers such as the nylons[57] and ethyl cellulose[58, 59], this component
can apparently by removed by annealing. It is interesting that with some glassy polymers the dual mode type of isotherm is not found. Presumably the sites are filled at
low relative vapor pressures and are masked by the regular swelling type of sorption.
Highly hydrophilic polymers such as cellulose swell so much that water clusters appear to exist as free water as shown for example by NMR and Infra Red measurements. The theoretical and practical aspects of water sorption has been described in
a number of papers[54, 60, 61] and will not be expanded on further in this review.

Transport and Sorption of Organic Compounds

In general, the diffusivities of penetrants that swell glassy and rubbery polymers
increase with concentration. The sorption isotherms are normally well-described
by the Flory-Huggins equation. Clustering of penetrant can also occur and cause
deviations from this behavior[62]. In the case of glassy polymers and strong swelling
solvents, so-called Case II transport can occur[63−79]. As shown in Fig. 6[70] an initial
linear increase in sample weight with time characterizes Case II uptake in film samples.
The linear dependence of uptake on time results from the existence of a clearly dis-
cernable propagating solvent-induced swelling front which moves through the polymer at a constant velocity. The polymer ahead of the front is largely free of penetrant,
while the polymer behind the front has essentially reached its equilibrium swelling
value corresponding to the temperature and pressure of the experiment. For Fickian

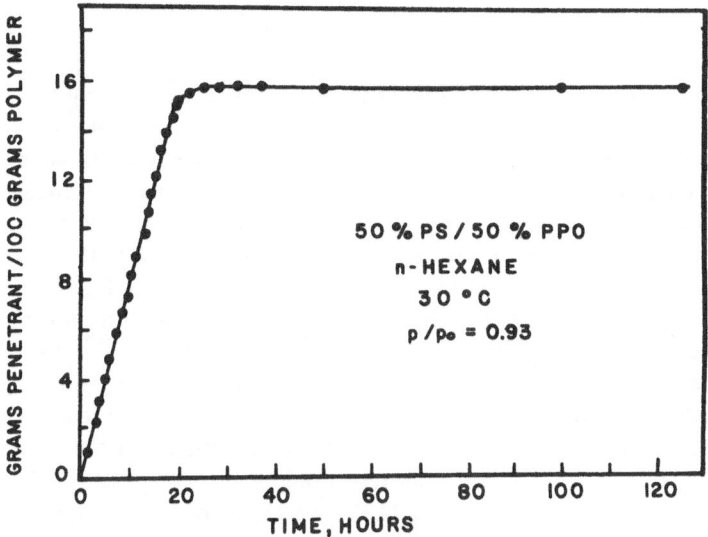

Fig. 6. Sorption kinetics of n-hexane in a 50/50 blend of polystyrene and polyphenylene oxide illustrating Case II sorption kinetics. Note that M_t/M_∞ is proportional to time rather than to the square root of time in the initial stages of sorption [70]

systems, the initial weight gain is linear with the *square root* of time; furthermore, a sharp advancing boundary separating swollen and unswollen region is usually not obvious.

Although the fundamental processes involved in Case II transport are not completely understood at the present time Sarti[80] has suggested that the resistance to advancement of the Case II solvent front is similar in magnitude to the resistance of the polymer to mechanical deformations. This resistance resides in the difficulty of initiating large scale segmental motion of chains in the glassy state. It is suggested that the rate of propagation of the Case II solvent front is essentially equivalent to the rate of craze propagation due to mechanical stress under the same effective stress conditon. In the Case II situation, stresses are exerted on the polymer as a result of the frustrated tendency of the glass to swell to its equilibrium value characteristic of the condition which eventually obtains in the swollen gel behind the advancing front. The sorption kinetics are limited by the relaxations of the chains to their swollen state since transport of solvent through the gel phase is rapid until the gel layer thickens considerably. Eventually, the uptake varies as t^a, where $0.5 < a < 1$. As the diffusional resistance of the gel phase increase, the exponent approaches 0.5. A considerable body of sorption kinetics data exists for cases in which both diffusion and relaxation contribute measurably to the transport mechanism[81-84]. Berens and Hopfenberg have presented a useful phenomenological treatment of sorption kinetics when both diffusion and relaxation contribute to the transport mechanism[85].

The sorption of organic vapors at high activities has been reviewed in recent papers[86-88]. At low concentrations, dual mode sorption takes place in glassy polymers as described later in this review. The effects of crystallinity and variable morphologies on gas and vapor transport in polymer blends will not be discussed here but an excellent review may be found in the literature[89].

Gas Separations

Commercial scale gas separations using membranes have been discussed actively for a number of years[90−96]. The "energy crisis" has sparked active interest in replacing the energy intensive cryogenic process currently used for most gas separations. Membrane processes have the potential for energy efficient operation and can provide an attractive solution to the problem of gas separation if selective membranes can be produced with sufficient surface areas to provide high throughputs. Work in this field has traditionally concentrated on the separation of He from CH_4 and N_2 from O_2; however, the "energy gases" CO and H_2 from coal gasification and oil reforming have been recent targets for the new generation of gas separation devices[96, 97]. Important developments in hollow fiber and asymmetric membrane technology now promise that large scale gas separations by membrane processes will become realities.

To be economically attractive, membranes used for gas separations must allow a high flux of the desired component while greatly restricting the passage of other species. The separation factor, $\bar{\alpha}$, defined in Eq. (13) is a measure of the selectivity of a membrane for some desired component "A" relative to some undesired component "B"[97]:

$$\bar{\alpha} = \frac{[Y_A/Y_B]}{[W_A/W_B]} \tag{13}$$

where Y_i is the mole fraction of the ith component in the downstream receiver and W_i is the mole fraction of the ith component in the high pressure upstream supply line. In 1866, Graham[1] showed that air could be enriched in O_2 by selective permeation through a natural rubber membrane. His reported enrichment of the downstream gas to 41% O_2 corresponds to a separation factor of 2.61, which is in fair agreement with more recent measurements[8]. Graham described the separation process qualitatively by a solution-diffusion model. He suggested that gases condense at the upstream face of the membrane and are transported at different rates through the rubbery matrix in the same fashion that gases diffuse through liquids by Brownian motion. For an ideal system such as gases in rubbers, Henry's law and Fick's law apply, and the gas diffusivity is a constant over a considerable range of pressures at a given temperature[8]. As noted earlier, von Wroblewski[3] showed that for such a case, the steady-state permeability of a gas is given by:

$$P = k_D D$$

where k_D is the Henry's law constant and D is the constant diffusion coefficient of the gas in the rubber. In such ideal systems, if the upstream pressure is significantly greater than the downstream pressure, the separation factor in Eq. (13) can be approximated by:

$$\bar{\alpha} \simeq P_A/P_B \tag{14}$$

where P_i is the steady state permeability of the ith component[97, 98].

Table 2. Rubbery polymer permeabilities and selectivities for various gases[d]

Rubber	T°C	P_{He} (Barrer)[a]	P_{H_2} (Barrer)[a]	P_{CO_2} (Barrer)[a]	P_{CH_4} (Barrer)[a]	P_{O_2} (Barrer)[a]	P_{N_2} (Barrer)[a]	P_{He}/P_{CH_4}	P_{O_2}/P_{N_2}	P_{H_2}/P_{CH_4}	P_{CO_2}/P_{CH_4}
Silicone rubber	25	300	550	2700	800	500	250	0.38	2.0	0.69	3.38
Silicone/poly-carbonate block copolymer	25	–	210	970	–	160	70	–	2.29	–	–
Silicone/nitrile copolymer	25	86	123	670	–	85	33	–	2.58	–	–
Natural rubber	25	31	49	131	30	24	8.1	1.03	2.96	1.63	4.4
SBR	25	23.1	40.1	124	21	17	6.3	1.09	2.71	1.91	5.9
Acrylonitrile/butadiene co-polymer (27% AN)	25	12.2	15.9	30.9	–	3.84	1.1	–	3.63	–	–
Acrylonitrile/isoprene (26% AN)	25	7.8	7.5	4.3	–	0.85	0.18	–	4.72	–	–
Butyl rubber	25	8.42	7.2	5.18	0.78	1.30	0.33	10.67	3.96	9.17	6.6
Polychloroprene	23	13	20	22	2.6	4.0	1.1	5.0	3.64	7.69	8.5
Polyethylene $\rho = 0.914$	25	4.9	–	12.6	2.9	2.9	0.97	1.69	2.99	–	4.3

Table 2 (continued)

Rubber	T °C	P_{He} (Barrer)[a]	P_{H_2} (Barrer)[a]	P_{CO_2} (Barrer)[a]	P_{CH_4} (Barrer)[a]	P_{O_2} (Barrer)[a]	P_{N_2} (Barrer)[a]	P_{He}/P_{CH_4}	P_{O_2}/P_{N_2}	P_{H_2}/P_{CH_4}	P_{CO_2}/P_{CH_4}
Polyethylene $\rho = 0.964$	25	1.14	–	1.7	0.39	0.41	0.14	2.90	2.93	–	4.4
FEP copolymer[e]	30	62	14.1	–	1.4–23[b]	5.9	2.2	44–2.7[b]	2.68	10.1–0.6[b]	–
Teflon[e] 75% crystalline	30	11	12[c]	2.6	–	–	0.79	–	–	–	–
Polyethylene/vinyl acetate copolymer	25	30	–	25	30	–	–	1.00	–	–	0.8
Chlorosulfonated polyethylene	23	7.2	10.8	15.8	1.7	2.1	0.92	4.23	2.28	6.36	9.3
Nylon 66	25	–	1.0	0.17	–	0.034	0.008	–	4.25	–	–

[a] 1 Barrer = 10^{-10} [cm³ (STP) cm]/[cm² sec cm Hg].

[b] Highly pressure dependent at low temperature. Above 60 °C, pressure dependence diminishes greatly[94, 100].

[c] Temperature = 33 °C.

[d] The above table was compiled from the large body of permeability data polymers reported in Refs.[7, 8, 23, 99, 101].

[e] These fluorocarbon polymers are difficult to unambiguously describe as rubbery or glassy due to multiple thermal transitions.

As the molecular size of a penetrant increases, the diffusivity decreases; however, since larger molecules tend to be more condensible, k_D increases with increasing penetrant size. Since one of the factors in the permeability equation tends to increase and one tends to decrease with increasing penetrant size, the permeabilities of very small molecules such as helium and quite large molecules such as CO_2 are often surprisingly similar in many rubbery materials. This compensating effect of molecular size on permeability reduces inherent membrane selectivity. This effect is obvious from the various permeability ratios presented in Table 2; attractive separation factors of 10 or greater are rather rare. One apparent exception to this statement is the system He/CH_4 in FEP copolymer[94]. The ratio of permeabilities of the two pure components at subatmospheric pressures is 44. When the methane driving pressure rises to 65 psi, however, the permeability of CH_4 rises rapidly to a value similar to that for helium in FEP at 30 °C. This loss in membrane selectivity at high driving pressures is reportedly not a problem above 60 °C[94, 100].

A proposal to amplify the separation factor of a material such as polyethylene for two gases such as H_2 and CH_4 in which the permeabilities are not strong functions of pressure[102] (see Table 3) entails operation in a pulsed or cyclic pressure mode with the permeate being collected alternately in two receiving volumes[103, 104]. For such a binary system in which $D_A > D_B$, the mole fraction of component A collected in the receiving volume which is open during application of the high upstream pressure pulse will be larger than for steady state permeation. Conversely, the mole fraction of component A collected in the second receiving volume (open during the low pressure portion of the pressure period) will be smaller than the value for steady state permeation. This approach is based on the fact that although the steady state transport of a gas in a rubber is governed by the product of k_D and D, the time interval required to build up to a steady permeation rate after raising or lowering the driving pressure is governed only by the gas's diffusion coefficient. Optimization of the pressure cycle period can provide very large increases in the separation factor, $\bar{\alpha}$, for gases with significantly different diffusivities. Reduced productivity associated with the low pressure portion of the cycle and the mechanical complexities associated with operation in such a cyclic mode may restrict application of this novel approach.

Glassy polymers are inherently more selective on the basis of penetrant molecular size and shape than are rubbers. The exact explanation of this phenomenon is possibly quite complex, but it is undoubtably related to the significantly lower

Table 3. Polyethylene permeability to H_2 and CH_4 at 30 °C

p psia	Permeability of H_2 $\dfrac{cc(STP)\ cm}{cm^2\ sec\ cm\ Hg} \times 10^{10}$	Permeability of CH_4 $\dfrac{cc(STP)\ cm}{cm^2\ sec\ cm\ Hg} \times 10^{10}$
40	10	3.26
115	10	3.24
315	9.7	3.33

backbone mobility in the glassy state. Tables 4 and 5 present some impressive permeability ratios for the systems: H_2/CH_4, H_2/CO, He/CH_4 and CO_2/CH_4 in a number of glassy polymers. The first three entries in Table 4 indicate that stiff-chained polyimides may provide the most selective gas separation membranes currently available. The major disadvantage of using glassy membrane materials for gas separations is their extremely low permeability.

The development of asymmetric membrane technology in the 1960's was a critical point in the history of gas separations. These asymmetric structures consist of a thin $(0.1 \mu$ to $1 \mu)$ dense skin supported on a coarse open-cell foam structure. Asymmetric membranes composed of the polyimides discussed above can provide extremely high fluxes through the thin dense skin, and still possess the inherently high separation factors of the basic glassy polymers from which they are made. In the early 1960's, Loeb and Sourirajan[107, 108] described techniques for producing asymmetric cellulose acetate membranes suitable for separation operations. The processes involved in membrane formation are complex. It is believed that the thin dense skin forms at the

Table 4. Glassy polymer permeability ratios for H_2 relative to CH_4 and CO at 30 °C

Material	$P_{H_2} \times 10^{10} \dfrac{cm^3(STP)\,cm}{cm^2\,sec\,cmHg}$	P_{H_2}/P_{CH_4}[l]	P_{H_2}/P_{CO}[l]	Ref.
3,5 DBA-6F[a]	33.6	340.6	–	105)
1,5 ND-6F[b]	78.1	108.5	–	105)
PPD-6F[c]	48.2	152.5	–	105)
Polystyrene	11.0	–	12	97)
PVC	8.0	–	12.9	97)
Parylene N[d]	2.8	–	25.4	97)
Cellulose Acetate[e]	13.0	–	37.2	97)
Sulfone[f]	14.0	–	37.8	97)
Polyvinyl fluoride[g]	0.6	–	66.7	97)
Mylar Type S[h]	1.4	–	74.0	97)
Polyimide[i]	2.0	–	74.0	97)
Parylene C[j]	1.4	–	110.0	97)
Polycaprolactam[k]	1.5	–	115.0	97)

[a]   Soluble polyimide prepared by reaction of 4,4'-hexafluoroisopropylidene diphthalic anhydride (6 F) with 3,5-diaminobenzoic acid.
[b]   Soluble polyimide prepared by reaction of (6 F) with 1,5-diaminonaphthalene.
[c]   Soluble polyimide prepared by reaction of (6 F) with 1,4-diaminobenzene.
[d]   Poly(para-xylylene) Union Carbide Reg. trademark.
[e]   Dupont 100 CA-43.
[f]   Union Carbide Sulfone 47.
[g]   Dupont Tedlar.
[h]   Dupont semicrystalline poly(ethylene terephthalate) film.
[i]   Dupont Kapton polyimide.
[j]   Poly(monochloro-paraxylylene) Union Carbide Reg. trademark.
[k]   Allied Chemical, amorphous nylon film.
[l]   Possibly dependent on driving pressure.

Table 5a. Glassy polymer selectivities for He and CO_2 relative to CH_4

Polymer Structure	P_{He} (Barrer)[b]	P_{CO_2}[c] (Barrer)[c]	P_{He}/P_{CH_4}[c]	P_{CO_2}/P_{CH_4}
Poly(tetramethyl bis L terephthalate)	148	83.0	16	34
Poly(tetra isopropyl bis A-sulfone)	110	98.0	25	15
Poly(2–6-dimethyl-1,4-phenylene oxide)	60.0	50.0	27	18
Poly(tetramethylcyclo-butane diol carbonate)	75.0	32.0	38	14
Poly(tetramethyl bis A carbonate)	43.0	29.0	49	33
Poly(tetramethyl bis L sulfone)	48.3	65.0	28	24
Poly(tetramethyl bis A isophthalate)	31.0	–	60	–
Poly(tetramethyl bis A sulfone)	38.4	21.0	67	68
Poly(bis A carbonate) Lexan	15.0	10.0	35	20
Poly(bis A-4,4'-N-methyl phenyl sulfone urethane)	6.05	–	93	–
Poly(ethylene terephthalate)	1.30	–	57	–

[a] No temperature was reported in the reference, presumably, these data apply at $\sim 30\,°C$[106].
[b] 1 Barrer $= 10^{-10}$ $cm^3(STP)\,cm/(cm^2\,sec\,cmHg)$.
[c] Likely to be somewhat pressure dependent.

interface between the casting solution and the air as a result of rapid evaporation of solvent prior to the final coagulation step. The coarse, open cell foam support is formed during the final coagulation step[109–112].

Kimura and Sourirajan[113] have offered a theory of preferential adsorption of materials at interfaces to describe liquid phase, selective transport processes in porous membranes. Lonsdale et al.[114] have offered a simpler explanation of the transport behavior of asymmetric membranes which lack significant porosity in the dense surface layer. Their solution-diffusion model seems to adequately describe the cases for liquid transport considered to date. Similarly gas transport should be describable in terms of a solution-diffusion model in cases where the thin dense membrane skin acts as the transport moderating element.

One can define the local permeability, P_i, of the ith species in terms of the flux of that species, J_i as shown by Eq. (15):

$$J_i = P_i \frac{\Delta p_i}{\Delta L} \qquad (15)$$

where Δp_i is the partial pressure difference in the gas separator of the ith component between the high pressure and low pressure side of the membrane and ΔL is the thickness of the thin dense skin which moderates the transport process. As discussed in the section concerning theories of transport of gases, both the "effective" solubility coefficient (C/p) and the diffusion coefficient, D_{eff}, can be significant functions of the local permeate concentration in glassy polymers. The dual mode sorption and partial immobilization transport models[39], discussed previously for pure gases, are necessary for a fundamental description of the transport processes in these systems based on glassy polymers.

The data of McCandless in Table 6 show that the actual separation factor at 83 ° C for the binary system H_2/CO in glassy Kapton polyimide film is less than that predicted from the ratio of the permeabilities of pure H_2 and pure CO[97]. Moreover, it was reported that the overall flux was 20–30% below that which one would predict using the pure component permeabilities. McCandless concluded that these data indicate "there is some interaction between components of the binary mixture during permeation which effectively decreases the permeation of H_2 relative to CO"[97]. In glassy polymers there may be preferential filling of available Langmuir sites by the more condensible species, and it is not surprising that the permeation of one species significantly affects the permeation of another species.

Pye et al.[102] report pressure dependent oxygen permeabilities in Mylar Poly-(ethylene terephthalate) and pressure dependent methane permeabilities in a glassy aromatic polyimide. Hydrogen permeability in the same polyimide membrane was essentially independent of pressure. The trends in these data, summarized in Table 7, are consistent with a study of various gases in glassy polycarbonate by Koros et al.[39]. The data in this later study were successfully analyzed in terms of the dual mode sorption, partial immobilization theory. The pressure dependency of the permeabilities and the reported interactions between permeating components noted above suggest that care should be exercised in basing design decisions for gas separators on pure component permeation data collected at a single measurement pressure.

The capability to mass-produce uniform, small diameter $(40-100 \mu)$ hollow fibers has been another important element in the commercial development of gas separators. Apparently, hollow fibers were first produced in the 1920's[93]; however, mass production of uniform fiber contours dates from the 1960's. Even with the

Table 6. Effect of upstream pressure on selectivity for hydrogen relative to carbon monoxide in glassy polyimide film

Upstream pressure (psi)	Actual separation factor $[Y_{H_2}/Y_{CO}]/[W_{H_2}/W_{CO}]$	P_{H_2}/P_{CO}
50	21.8	54.1
500	31.2	44.0

Notes: 50/50 mix on supply side, 1 atm permeate pressure in down-
stream receiver.
Temperature = 83 °C.
Kapton polyimide film, Dupont Reg. trademark.

Table 7. Effect of upstream pressure on the permeability of methane and oxygen in glassy polyimide and glassy Mylar

Upstream pressure (psia)	Polyimide Methane permeability (cB)	Mylar Oxygen permeability (cB)
115	43.2	3.5
315	36.9	–
615	30.4	–
1015	28.5	2.6

Note: $1 \text{ cb} = 10^{-12} \dfrac{\text{cm}^3(\text{STP})\,\text{cm}}{\text{cm}^2 \text{ sec cm Hg}}$.

high surface-to-volume ratios characteristic of hollow fibers, the transport resistance offered by a 10–20 μ wall thickness of glassy membrane in the totally dense morphology is too large to provide sufficiently high fluxes to be attractive for commercial operations. When the hollow fiber and glassy asymmetric technology are combined, however, a highly attractive candidate for commercial operation emerges. Presumably, new devices for gas separations will be based primarily on small diameter, glassy asymmetric hollow fibers.

A recent article[115] presents case studies of applications of hollow fiber permeators for gas separations in ammonia synthesis, oxo-alcohol synthesis and natural gas purification. In typical permeators a bundle of as many as 50 million fibers[93] is epoxied into tube sheets and encased in a pressure vessel with suitable fittings for feed and product flows. The feed gas enters the unit and is distributed by the tube sheet to the fibers. As the feed gas flows inside the fiber, the various species transport to the shell side according to Eq. (15). Processing rates of 2.25×10^6 SCFD per permeator are reported for the DuPont design in H_2/CO applications[115]. These permeators are roughly 15 feet long by one foot in diameter[116]. Various modes of operation of the permeators are described by Atonson[116] et al. Pressure dependence of the apparent solubility and diffusivity were acknowledged to be important in certain cases; however, it was stated that in most situations adequate design results could be obtained by using constant "effective" values for k_D and D over the pressure range of interest[116].

A commercially available alternative to the use of glassy asymmetric membranes has been pioneered by General Electric in the form of block copolymers of silicone rubber and polycarbonate[99]. Hollow fibers of pure silicone rubber are highly permeable to many gases but tend to distend under the influence of high internal driving pressures. The block copolymer provides a novel solution to this mechanical stability problem. In this approach, the inherently high fluxes characteristic of silicone rubber are maintained, while the polycarbonate blocks serve as physical crosslinks to make the membranes mechanically self-supporting in the hollow fiber configuration. The polycarbonate composition must be kept low (<20%) to avoid formation of complex lamellar and rod-like structures in the dispersed phase[89] which could interfere significantly with the permeation process. In the ideal case where the polycarbonate islands act strictly as crosslinks the membrane selectivity should be essentially that of silicone rubber. The effect of the presence of the dispersed polycarbonate phase on the steady state permeability in these systems should be treatable in terms of transport theories of composites reviewed by Hopfenberg and Paul[89]. Design of hollow fiber permeator units using these copolymers according to the principles of Antonson et al.[116] should be acceptable.

Water Purification

The use of synthetic polymeric membranes for water purification is now an established technology. Historically, this development dates to the beginning of this century, when Zsigmondy and Bachmann[117] prepared the first microporous membrane from cellulose esters. Similar microfiltration membranes are now widely used in applications ranging from sterile filtration to fine particle removal.

In recent years, two other important applications of synthetic polymeric membranes in water purification have become established. Chronologically the first of these, reverse osmosis, is rapidly becoming the principal method of water desalination worldwide. Another membrane process, ultrafiltration, is even newer and is finding important use in the removal of high molecular weight, colloidal, and emulsified materials from aqueous streams.

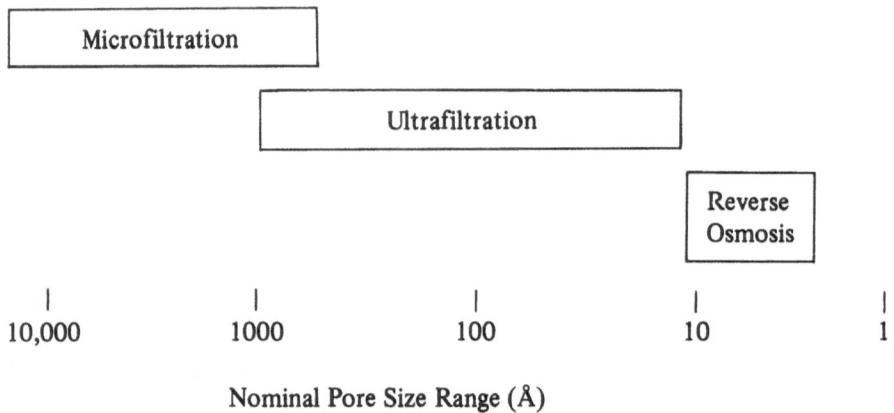

Nominal Pore Size Range (Å)

Microfiltration:	Particles down to 500 Å diameter. Low pressures, 5 to 50 psi. Very high specific fluxes, >1 m^3/m^2-day-atm.*
Ultrafiltration:	Particles from 1000 to 10 Å diameter. Moderate pressures, 15 to 100 psi. Moderate specific fluxes, 0.1–0.5 m^3/m^2-day-atm.
Reverse Osmosis:	"Particles" from 10 to 2 Å diameter. High pressures, 100 to 1500 psi. Low specific fluxes, <0.05 m^3/m^2-day-atm.

---

* 1 m^3/m^2-Day = 24.6 gal/ft^2-day.

While these membranes tend to be classified according to the pressure-driven process in which they are used-microfiltration, ultrafiltration, and reverse osmosis (or MF, UF, and RO) — all of these membranes actually from a continuum in properties, from filtration barriers with well-defined pore sizes to non-porous membranes that effect separations on the basis of the chemical properties of the permeants rather than their physical size. The three types of membranes are also usually made, in fact, by similar processes, all involving controlled precipitation of a polymer/ solvent casting solution by water. With MF membranes, precipitation is achieved by inhibition of water from a humid atmosphere[118]. This process is relatively slow and, as a consequence, rather open porous structures are formed. In the Loeb-Sourirajan[107, 108] process widely used for preparing cellulose acetate RO and UF membranes, the freshly cast film is immersed in a water bath. Rapid precipitation occurs in this case, and a finely porous membrane structure is produced, usually with a very thin and relatively dense upper surface or "skin" where the precipitation is most rapid. Kesting[119] and Strathmann et al.[120] have described in detail the methods by which these membranes are prepared, emphasizing the similarities between the methods. A summary of the range of "nominal pore sizes" of these three classes of membranes and the typical operating conditions and specific water fluxes is presented above (see p. 92).

In this section, we describe recent developments in water purification membranes, beginning with RO membranes, which are the "tightest" and most permselective, and proceeding to the more open UF and MF membranes. We close with a short discussion of liquid membranes, which are generally not polymeric membranes but which merit consideration here because they are highly permselective, they are potentially useful in water purification, and they process certain unique and interesting properties.

A. Reverse Osmosis

Water can permeate through non-porous membranes by a solution-diffusion process. The driving force for permeation is the net pressure difference, $\Delta p - \Delta \pi$, and the water flux can be represented by the equation[114]

$$J_1 = \frac{D_1 C_1 \overline{V}_1 (\Delta p - \Delta \pi)}{RT \, \Delta X} \tag{16}$$

where D_1 is the diffusion coefficient of water and C_1 is the solubility of water in the membrane, $\overline{V}_1$ is the partial molar volume of water, Δp is the applied pressure difference and $\Delta \pi$ is the osmotic pressure difference across the membrane, R and T have their usual significance, and ΔX is the effective membrane thickness.

Salt can also permeate by a solution-diffusion process, but the driving force for salt permeation is the concentration difference rather than the net pressure difference. The relationship governing salt flux is[114]

$$J_2 = \frac{D_2 K \Delta C_2}{\Delta X}, \tag{17}$$

where K is the membrane-solution partition coefficient for salt and ΔC_2 is the external salt concentration difference. Thus, water flux increases with increasing pressure but the salt flux does not, and the quality of the permeate stream improves with increasing applied pressure. A widely used index of permeate water quality is the salt rejection, defined as

$$S = (C_2' - C_2'')/C_2' = \Delta C_2/C_2' \tag{18}$$

where the superscripts ' and " refer to the feed and permeate solutions, respectively. We can substitute from Eqs. (16) and (17) to obtain

$$S = \left[1 + \frac{D_2 KRT}{D_1 C_1 \overline{V}_1 (\Delta p - \Delta \pi)} \right]^{-1} \tag{19}$$

Research and development in RO membranes has centered on a) finding materials with high water permeability ($D_1 C_1$) and low salt permeability ($D_2 K$), in order to optimize water flux and salt rejection; b) making membranes that are effectively very thin (i.e. small ΔX), in order to maximize water flux; and c) packaging membranes into inexpensive modules with high surface area per unit volume, in order to decrease capital costs. In fact, two distinct approaches to b) and c) have been followed. On the one hand, imperfection-free membranes of extreme thinness (effectively, $\Delta X \simeq 1000$ Å or even less) have been developed, exhibiting water fluxes approaching 1 m^3/m^2-day with high salt rejection. These flat sheet membranes have been incorporated into inexpensive spiral wound modules[121]. In a separate approach to low cost desalination systems, membranes have been prepared in the form of very fine hollow fibers[122]. Although these exhibit relatively low water fluxes (0.1 m^3/m^2-day or even less), they can be incorporated into inexpensive modules with packing densities approaching 75,000 m^2/m^3. Interestingly, the two approaches are commercially quite competitive.

The search for polymeric materials with favorable transport properties has been entirely empirical in the past. Two materials are in common use today: cellulose acetate, with a degree of acetylation ranging from 2.5 to 2.8[114], and a family of aromatic polyamides and polyamide-hydrazides[123]. Both types of materials are moderately hydrophilic. The cellulose acetates sorb 10—15 wt% water at 100% relative humidity, and have a water permeability of approximately 10^{-7} g/cm-sec. Their permeability to NaCl in dilute aqueous solution is about three orders of magnitude lower, and it is generally true that effective RO membranes have a water-to-salt permeability ratio, $D_1 C_1 / D_2 K$, in the order of 10^3 or more. The aromatic polyamides exhibit similar water sorptions, but the salt diffusivity is lower than that in cellulose 2.5-acetate[124]. The aromatic polyamides have a stiffer backbone than the cellulose acetates and this apparently results in reduced mobility of permeants. While we address below recent developments in RO membranes, we should emphasize at this point that the dominant RO membranes in use today are the Loeb-Sourirajan type cellulose acetate membranes in spiral-wound module form and the aromatic polyamide hollow fine fiber membranes packaged into a multiple tube-in-shell design.

RO membranes depend for their performance on large differences in their permeability to water and salt. These permeabilities are determined by two factors: diffusivity and solubility (or partition coefficient). These two parameters are not totally unrelated. Materials with higher water sorbtivity are also more highly plasticized by this water and can be expected to exhibit higher diffusivities as well. However, none of the effective RO membrane materials are highly hydrophilic because salt permeability tends to increase much more rapidly than water permeability with increased hydrophilicity. The best membranes also sorb water without permitting the water molecules to cluster. That is, large aqueous domains, which could dissolve salt, are avoided and the salt solubility in the membrane, when expressed as grams salt/gram sorbed water in the membrane, is generally well below the aqueous solubility. The number of papers in which diffusivities and partition coefficients for water and salt are reported is still quite limited. However, it is clear that the permselectivity of RO membranes derives principally from differences between salt and water diffusion coefficients rather than differences in partitioning. Thus far, our knowledge ends here; that is, good correlations do not exist between micromechanical or chemical properties of polymers and salt and water transport properties. We are a long way from having a predictive model.

In recent years, RO membrane research has proceeded in two directions. First, there has been a continuing search for new polymeric membrane materials. Some of the materials with interesting properties that could be cited include other cellulose esters[125], polybenzimidazole[126], a polybenzimidazolone[127] (PBIL), poly-imides[128, 129], and new aromatic polyamides[130, 131].

A summary of the properties of some of these materials is presented in Tables 8 and 9. Diffusion coefficients and NaCl partition coefficients are presented in Table 8 for cellulose triacetate and for Nomex® polyamide. In most cases, however, intrinsic transport properties are not known, and what is reported in the literature is the water flux and salt rejection of RO membranes under given test conditions. Under comparable test conditions (0.5–1% NaCl, 27–68 atm) the water fluxes of these newer membranes are comparable (0.4–0.8 m^3/m^2-day). Salt rejection data are presented in Table 9. To put these data in perspective, the salt rejection of cellulose 2.5-acetate membranes is typically <99%. However, for various reasons none of these new membranes has yet become commercially important.

The second current research direction is concerned with new methods of preparing extremely thin membranes. Typically, these membranes are less than 1 μm in effective thickness, and a film this thin requires some sort of support to withstand

Table 8. Diffusion coefficients and NaCl partition coefficients for cellulose triacetate and Nomex polyamides

Membrane	D_1 (cm²/sec)	C_1 (g/cm³)	D_2 (cm²/sec)	K
Cellulose triacetate, 43.2% acetyl[114]	1.3×10^{-6}	0.12	3.9×10^{-11}	0.015
Nomex® polyamide[124]	1.5×10^{-6}	0.2–0.49	1.3×10^{-10}	0.2

Table 9. Salt rejection data

Membrane	NaCl rejection (%)	Ref.
Cellulosics		
Cellulose triacetate	99.2	114)
Cellulose acetate blend	99.4–99.8	132)
Cellulose acetate methacrylate	99.7	125)
Cellulose acetate propionate	99.5	125)
Cellulose acetate butyrate	99.4–99.8	125)
Polyamides		
Nomex® aromatic polyamide	99.5	128)
APl aromatic polyamide	99.8	130)
PA-6 polyamide hydrazide	99.8	128)
Fused ring heterocyclics		
Polybenzimidazole	99.0	126)
Polybenzimidazolone	99.0	127)
Polybenzimidazopyrrolone	99.8	128)
Aromatic polyimide	99.5	128)

the operating pressures typical of RO operation (20–50 atm). A schematic representation of the structure of a supported thin film is presented in Fig. 7. The microporous support films used in making these "composite" membranes have typically been prepared from polysulfone or cellulose esters. They are, in fact, similar to the microfiltration membranes discussed in Sect. C below.

A variety of methods have been explored for preparing the thin salt-rejecting skin on one surface of these finely porous supporting membranes. In the earliest method, a film of cellulose acetate was laid down by dipcoating the supporting membrane with a dilute solution of the polymer[133]. When the solvent evaporates, a thin film of polymer, tightly adherent to the supporting membrane, is left behind. Films as thin as a few hundred Angstrom units have been made in this way, without any apparent imperfections.

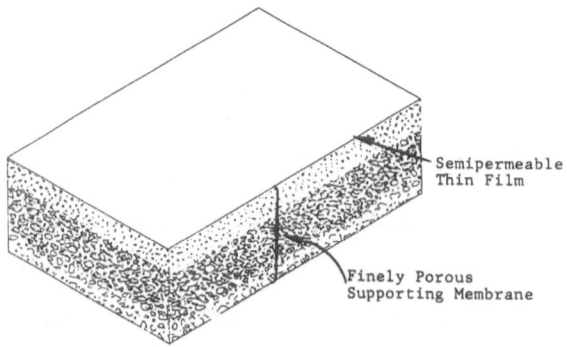

Semipermeable
Thin Film

Finely Porous
Supporting Membrane

Fig. 7. Schematic cross section of a composite membrane[134]

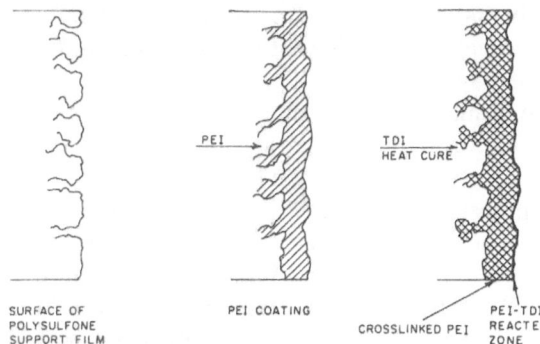

SURFACE OF   PEI COATING   PEI-TDI   **Fig. 8.** Schematic representation
POLYSULFONE   CROSSLINKED PEI   REACTED   of the NS-100 membrane[135]
SUPPORT FILM   ZONE

More recently, composite membranes have been made by interfacial polymerization or by *in situ* polymerization[134−136]. A representative case is illustrated in Fig. 8.
Here, a microporous polysulfone membrane is used as a substrate. This membrane is
soaked in a dilute aqueous solution of a low molecular weight polyethylenimine
(PEI). Without drying, this membrane is then contacted with a crosslinking agent
such as toluene diisocyanate (TDI) or isophthaloyl chloride dissolved in hexane,
after which the membrane is cured in an oven. A highly crosslinked, salt-rejecting
interfacial layer is formed in this way. A summary of the properties of three of the
more important composite membranes is presented in Table 10.

These composite membranes offer some distinct performance advantages over
the more conventional RO membranes, and there is growing interest in their further
development. Some large desalination plants are under construction embodying
these membranes.

Certain other approaches to preparing composite membranes have also been
examined, but with minimal practical realization thus far. "Dynamically formed"
membranes for RO and UF have been prepared by passing a colloidal suspension of,
for example, a hydrous metal oxide over a finely porous substrate[137]. A thin, salt
rejecting layer is deposited in the pores. Such membranes exhibit very high water
fluxes (up to 10 m^3/m^2-day) with moderate salt rejection. However, their properties
tend to be somewhat transitory and difficult to reproduce. "Plasma polymerized"
membranes have been prepared by creating an electrodeless glow discharge in a hydro-

Table 10. Composite membrane performance

Membrane	Type	Water flux (m^3/m^2-day)	NaCl rejection (%)	Ref.
NS-100	Crosslinked poly-ethylenimine	0.8−1.0	to 99.5	135)
NS-200	Crosslinked furfuryl alcohol	0.8	to 99.9	136)
PA-300	Crosslinked poly-(ether-amide)	0.8−1.0	to 99.4	134)

carbon vapor in the presence of a finely porous substrate[138]. Again, a very thin, salt-rejecting film is deposited, and excellent performance has been observed in RO. Reproducibility is also a problem in these systems.

One other form of pressure-driven membrane process should be mentioned: piezodialysis[139–142]. In this process, a selectively salt-permeable membrane is used and the application of pressure to the salt-containing solution forces a salt-enriched solution through the membrane. The process is similar to RO in that a high pressure must be applied, and the product purity increases with increasing applied pressure. (The "product" in this case is the non-permeating solution.) What is required, then, is a membrane in which the salt solubility, expressed as grams salt/gram water sorbed in the membrane, is much greater than the aqueous solubility. Salt-enriching "mosaic" membranes have been developed by creating alternately cation- and anion-permeable regimes in a membrane. Provided the spacing between these regimes is sufficiently fine, each ion type will permeate through its own region. Highly permselective membranes have been made in this way, the best of which exhibit substantial salt enrichment in the permeant stream. Nevertheless, piezodialysis has not been advanced to the stage of practical reality.

B. Ultrafiltration

In a sense, this is the newest of the pressure-driven membrane processes. Even though UF was practiced on a laboratory scale in the 1930s, it has become a *bona fide* industrial process only within the past decade.

The theory of UF is not sharply defined. Generally, a pore-model is used to describe these membranes, and it is recognized that all UF membranes exhibit a pore size distribution. Molecules larger than the largest pores are assumed to be completely rejected, and those smaller than the smallest pores will permeate rather freely. This still leaves a transition region in which partial rejection occurs. While it is conventional to speak of a molecular weight (MW) cutoff in UF membranes, i.e. a molecular weight above which molecules will be completely rejected, these cutoffs tend to be rather diffuse in real membranes. From a practical point-of-view, this limits the utility of these membranes in separating molecules on the basis of size. Generally, clean separations can only be achieved between molecules differing in MW by at least an order of magnitude. It should also be noted that straight chain polymeric molecules will permeate UF membranes relatively freely even though globular molecules of the same MW will be well rejected[143]. Thus, molecular shape can be important as well.

By definition, UF membranes are freely permeable to inorganic salts and other molecules with MW less than about 1000. Because it is these species that generally create most of the osmotic pressure of solutions, the net osmotic pressure difference across UF membranes is generally quite small and therefore small applied pressures can be used. Because these membranes are more open than RO membranes, there is less necessity to produce very thin membranes in order to achieve high water fluxes.

In practice, UF operation is usually limited by "cake formation"[143, 144]. In all pressure-driven membrane processes both the solvent (water) and solutes are carried convectively to the membrane surface. In UF, only the solvent and microsolutes pass through the membrane. The macromolecules are rejected and, because of their small

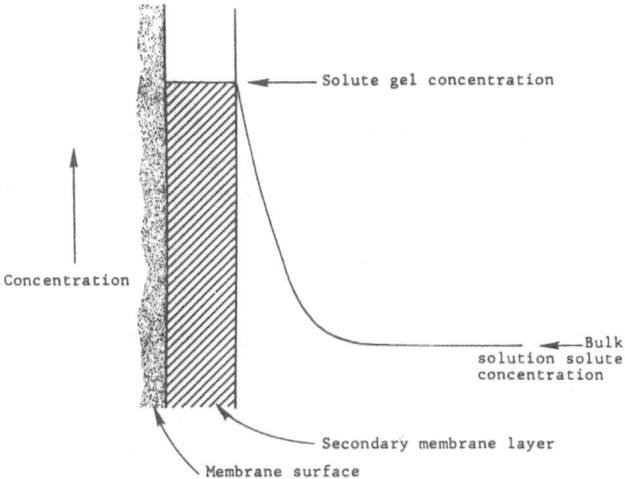

Fig. 9. Schematic representation of the concentration profile of rejected solute at the surface of an ultrafiltration membrane

diffusivity in solution, they tend to accumulate at the membrane surface, a phenomenon referred to as concentration polarization. Typically, macromolecular solutions exhibit a concentration which, if exceeded, will cause the solution to gel. Because of concentration polarization, this gelation point is always reached first at the membrane surface. The phenomenon is represented schematically in Fig. 9. The formation of this gel layer is usually referred to as "cake formation"; it amounts to a secondary membrane on the surface of the UF membrane. Once formed, it dominates the process. If the operating pressure is increased, the flux temporarily increases, causing the thickness of the secondary membrane to increase, and the flux again falls. The secondary membrane can also change the rejection characteristics or "molecular weight cutoff" of the membrane[143].

The principal current applications of UF are the separation of oil from emulsified oil wastes; recovery of paint from electro-coat paint rinse tanks; and the concentration of whey, a by-product in cheese making[145−147]. In all of these applications, the process is governed by cake formation. Under these circumstances, the only requirements of the UF membrane are:

(1) It should be resistant to a broad range of chemical environments (pH, oxidants, solvents, etc.) and temperature.

(2) It should be strong, inexpensive, and readily manufacturable into useful shapes such as tubes or small fibers.

The material in most general use in UF membranes is cellulose acetate. The membranes are made following the general procedure developed by Loeb and Sourirajan for casting cellulose acetate RO membranes. The MW cutoff can be varied by altering the membrane annealing conditions, or by the addition of swelling agents to the casting solution. Recently, other UF membranes have been introduced that exhibit greater chemical resistance than cellulose acetate. These include a polyvinyl chloride-polyacrylonitrile copolymer, polysulfone, and other materials. A summary of the properties of most of the important UF membranes is presented in Table 11.

Table 11. Ultrafiltration membranes

Membrane type	Advantages/disadvantages
Cellulose acetate	Low molecular weight cutoffs possible (<1000 MW). Not autoclavable; limited pH range (4−7); must be stored wet
Polysulfone	Autoclavable; wide pH range (0−14); molecular weight cutoff generally >10,000
Polyvinyl chloride-Polyacrylonitrile copolymer	Not autoclavable
Polyelectrolyte complex	Wide pH range (2−12). Not autoclavable; must be stored wet

C. Microfiltration

In many respects, MF is a simple extension of UF. MF, however, is a genuine filtration process in which particulate matter typically 0.1−10 μm is removed from aqueous streams. The membranes are totally non-selective with respect to osmotic-pressure-generating solutes, and low pressure operation is the norm. Particle removal is achieved strictly on the basis of size, and the rules of particle filtration, particularly surface filtration, apply.

From the point of view of water purification, one of the earliest applications of MF is still the most important: sterile filtration. Bacteria can be removed from drinking water, solutions for intravenous or parenteral injection, heat-sensitive liquid foods, and so on by MF. The earliest MF membranes were prepared from cellulose nitrate or from mixed acetate and nitrate esters of cellulose.

The scope of applications of MF has been broadened in recent years by the introduction of inert membranes, particularly polypropylene, polycarbonate, and Teflon®. In general, these materials cannot be made by the methods developed for the cellulosics because of their insolubility. Both Teflon® and polypropylene MF membranes have been made by a controlled stretching procedure in which microtears are introduced[148]. Microporous polycarbonate membranes have been prepared by a unique radiation-track-etch method[149]. A thin polycarbonate film is exposed to ionizing radiation which leaves labile sites that can later be chemically etched to produce straight-through channels. The pore size can be controlled by the etching conditions. The pores in these membranes, contrary to those in cellulosic membranes, are quite uniform in diameter.

D. Liquid Membranes

Traditionally, membranes have been known to be inefficient permselective barriers, for two reasons. First, diffusion coefficients in solid polymeric materials tend to be low. This means that fluxes also tend to be low, a problem only partly ameliorated by the development of very thin membranes. Second, membranes are not highly permselective, in general, to molecules of similar molecular size and similar chemistry. (For markedly dissimilar molecules, other separation methods are fequently straightforward and inexpensive.) That is, if a membrane is highly permeable to molecule A, it will not be highly impermeable to molecule B if A and B are chemically related.

A number of organic liquids are known, however, that exhibit extreme selectivities between chemically related species. Perhaps the best known examples are the crown ethers, which form stable complexes with potassium ion but not with sodium ion. Also important are the so-called liquid ion exchangers which can exchange metal ions for hydrogen ions or *vice versa,* depending on pH. Chemically, these agents are typically oximes, quinolines, or amines. They are frequently highly hydrophobic.

If such agents are incorporated into microporous membranes, they are held tightly by capillary forces, and thus create a liquid membrane. These membranes have interesting and unique properties. For example, they exhibit high permeabilities, because the diffusion coefficients are typical of those occurring in liquids (10^{-5}– 10^{-6} cm^2/sec at room temperature) rather than those occurring in solids (typically $<10^{-8}$ cm^2/sec). Second, they are highly permselective. Copper-iron separation factors of several thousand have been reported, for example. Finally, they can be used to pump ions up steep concentration gradients[150–153]. Consider Fig. 10. Depicted schematically there is a liquid ion exchange membrane. A gradient in hydrogen ion concentration and in the concentration of a divalent metal ion are impressed in the same direction across the membrane. At the lefthand membrane-solution interface, two hydrogen ions are released to the solution and a metal ion is taken up by the organic membrane, forming the neutral species R$_2$M. This species can diffuse to the opposite membrane-solution interface where the reverse reaction takes place: The metal ion is released and two hydrogen ions are taken up. The complexing agent, R, thus acts as a shuttle, carrying metal ions across the membrane in one direction and hydrogen ions in the opposite direction.

It can be shown that equilibrium is established when the concentration ratios are given by the expression[151]:

$$\frac{[M^{2+}]_o}{[M^{2+}]_\varrho} = \left(\frac{[H^+]_o}{[H^+]_\varrho}\right)^2 \tag{20}$$

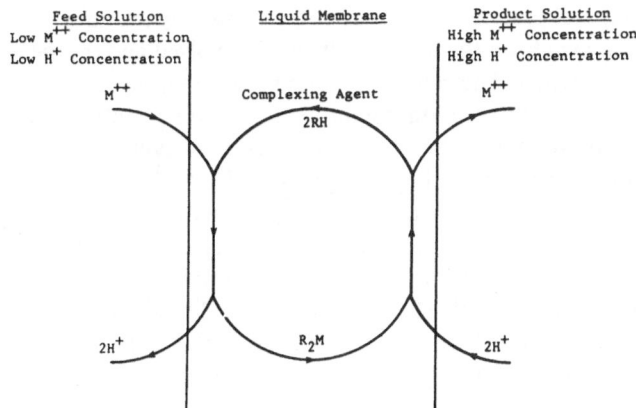

Fig. 10. Coupled transport of a divalent metal across a liquid membrane[151]

where o and ℓ refer to the two sides of the membrane. Thus, when presented with the proper pH gradient, metal ions will diffuse "uphill". With a pH difference of one unit, divalent metal ions can be concentrated 100-fold. With a pH difference of two units, divalent metal ions can be concentrated 10,000-fold. These predictions have been realized experimentally[151].

These membranes thus have some of the characteristics of biological membranes. Biomembranes are highly fluid and, because of their low viscosity, they exhibit high diffusivities. They are also highly permselective, although for different reasons. However, both types of membranes can behave as "chemical pumps".

This is the most recent development in membrane technology. It is currently being explored for a variety of applications, including recovery of metals from low grade ores by hydrometallury and recovery of metal ions from a variety of waste waters for pollution control.

Controlled Release

Much of the basic interest in the sorption and subsequent diffusion of various solutes in polymers has traditionally arisen from the need for barrier materials that exclude or contain substances to an adequate degree defined by the particular requirements of the applications. The ideal barrier would not transmit any of the gas, liquid, or solid solute in question, which, of course, in an absolute sense is an impossibility. Later, a renewed impetus for more fundamental insight into the sorption-diffusion characteristics of polymers was stimulated by the idea that semi-permeable membranes could provide useful and energy efficient separation devices. The ideal membrane would transmit one component at a very high rate while totally prohibiting the passage of another; however, the basic nature of these processes does not usually lead to such a circumstance, but many attractive compromises do exist. More recently, a different concept has provided even further reasons for interest in solute sorption and diffusion in polymers, and this is the idea of sustained or, hopefully, controlled delivery of a solute to achieve a particular objective. Here, the ideal does not involve absolute impermeability or infinite selectivity, but rather the challenge is to achieve a predetermined level and temporal pattern of transport. While this may sound simpler, it, too, places severe demands on the basic mechanisms involved and on the characteristics required of the polymer. This application has grown enormously in the last decade, and the discussion in this section, therefore, seeks only to introduce some of the more basic concepts and uses rather than to provide an encyclopedic review. Some recent publications better serve the latter purpose[154–166].

The basic idea of controlled release might be most readily appreciated by some oversimplified illustrations of the use of drugs to achieve desired responses in the body or to correct medical disorders[154, 167]. An example of the former is the use of well-known hormones for fertility control. The most widely used method of delivery is daily ingestion of a pill and the subsequent distribution of this drug throughout the entire body rather than directing it only to the necessary target, the fertility organs. Systemic distribution, in this case, is the probable cause of many adverse effects which are compounded by the larger quantity of the drug than actually required which must be consumed in this approach to delivery. Some approaches to controlled delivery resolve this situation by directing the drug to the sepcific target where its action is desired. Many medical disorders are treated by periodic injection or ingestion of drugs. At some time following this, the drug concentration in the blood stream rises to a high level and then falls gradually as the drug is processed by the body[167], and at some point the cycle is repeated. Within the same cycle, the drug concentration may become dangerously high or too low for therapeutic effectiveness. By controlled release methods of delivery, the drug concentration may be maintained effectively constant at an optimal level for prolonged periods.

While the applications of this concept are more dramatic, well developed, and widely publicized in the field of medicine, there are some equally important applications in other areas[168]. For example in agriculture, controlled release of pesticides[155, 156, 162], fertilizers[168, 169], and herbicides[170] offers similar advantages of more efficient application with less environmental side effects than conventional administrations since smaller amounts of the chemical agents must be delivered to achieve the desired effect. One of the first commercial applications of controlled release was in antifouling coatings for marine use[171]. Fouling of ship hulls increases fuel consumption[171] by as much as 20—25%, and the U.S. Navy is currently investigating antifouling paints that will reduce this problem without inflicting significant environmental damage. Products for control of household pests through sustained release of pesticides in small amounts have been available for some time.

A majority of the devices for controlled administration of chemicals or drugs employ polymers in their formulation to serve as a vehicle and/or rate controlling element. The purpose of the polymer and the mechanism by which it functions may vary in complexity, but clearly the sorption and diffusion of these chemicals in the polymer is frequently a central issue in product design and function. In addition, environmental stability and compatibility along with mechanical behavior and processability are also important issues. Further, these applications place unusual demands on quality control[172] of both materials and processes.

In the following we review some of the more important physical concepts, but not all, commonly employed in controlled release technology. In this discussion, we will point out some of the current applications and the important characteristics of the polymer.

Monolithic Devices

The simplest and most economical formulation to fabricate is to compound the agent of interest directly into a polymer matrix through conventional melt or solution mixing followed by molding, extrusion, or casting into the required geometrical shape, e.g. a platelet or pellet. After this device is placed in the desired location, the agent is released to the surrounding environment by any of several mechanisms. The details of the design and the operative mechanism will determine the rate of release and how this may vary in time.

The simplest case is when the agent is dissolved in the polymer matrix, i.e. its loading, A, does not exceed the solubility limit C_s in the matrix, and subsequent release is by simple molecular diffusion. Well-known diffusion equations[173–176] describe this process. During the early stages, the total amount of release, M_t, varies according to the well-known square-root of time relationship, i.e.

$$\frac{M_t}{M_\infty} = 2\,\frac{S}{V}\,\sqrt{\frac{Dt}{\pi}} \tag{21}$$

where S and V are the surface area and volume, respectively, of the device and D is the diffusion coefficient of the agent in the polymer. Clearly, such a device does not deliver at a constant rate and its ultimate capacity, M_∞, is frequently limited rather severely by the solubility of the agent in the polymer.

A more common situation is when $A > C_s$ and the bulk of the agent is physically dispersed in the matrix rather than dissolved in it. In general, release kinetics in this situation are complex, but in many cases the mechanism still involves a simple solution-diffusion process. The physical situation involves a core of suspended agent surrounded by a zone containing only dissolved agent, and this boundary advances inward as agent diffuses through this outer layer and leaves the device (see Fig. 11). Higuchi[177, 178] has developed a useful pseudo steady-state analysis for planar geometrics that predicts the following release pattern

$$M_t \simeq S\,\sqrt{2\,DC_s(A - 1/2\,C_s)\,t} \tag{22}$$

which differs only slightly from more rigorous analyses[179]. Once again, a M_t vs $\sqrt{t}$ pattern of release is predicted. These analyses have also been extended to other geometries[178, 180] and to include boundary layer resistances[179, 181]. It is an inherent feature of most shapes to result in a declining release rate with time[182]; however, it is possible to construct complex shapes[183] that give a nearly constant rate of release, frequently referred to as zero order release. Such designs result in a near

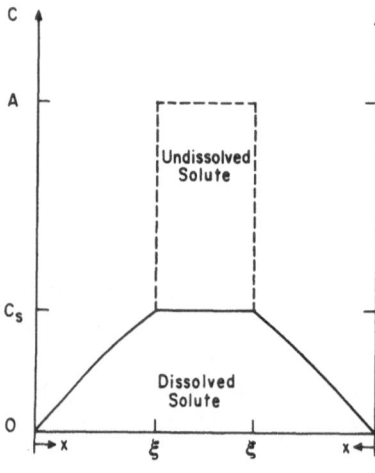

Fig. 11. Schematic profile of dissolved and undissolved solute concentrations for Higuchi model (courtesy J. Membrane Sci.)

compensation of the increased diffusion path as the depleted zone advances inward by an increase in area of this boundary. In any case, the important physical parameters are the solubility, C_s, and diffusivity, D, of the agent in the polymer. For a given agent, these can vary widely depending on the polymer choice; however, the challenge is to achieve the desired transport behavior simultaneous with other requirements on mechanical properties, stability, compatibility, and processing considerations. Several commercial devices[155, 156, 163] based on this concept are available for control of fouling, mollusks, and household insects. The literature contains many reports[154, 163, 167] of studies of such devices for medical purposes, e.g., birth control, treatment of hypertension, etc.

Generally, the agent to be released is a relatively small molecule with a molecular weight no larger than a few hundred. One would not expect that macromolecules, e.g. proteins, could be released by such a technique because of their extremely small permeation rates through polymers. However, Folkman and Langer[184] have reported some surprising results that clearly demonstrate the opposite. These results force one to consider alternate mechanisms than that of the classical Higuchi-type picture. Certainly, at high loadings, one could develop interconnecting channels through which macromolecules can diffuse more readily than the polymer matrix, and this might explain the interesting and important observations reported by Folkman[184].

Similarly, McGinnity[185] has shown that incorporation of hygroscopic "carriers" such as sodium alginate into silicone rubber based matrix devices greatly increased the rate of morphine sulfate release and changed its temporal character to approximate more nearly a zero-order pattern. It was noted that inclusion of a "carrier" caused the device to swell considerably because of water imbibition and no doubt a complex series of events account for the effect on release rate.

The environmental fluid surrounding the matrix devices considered here, usually water, can bring about changes in the polymer vehicle to cause it to release the agent dispersed within it. For example, the polymer might be slowly dissolved or chemically eroded away perhaps by hydrolysis, leaving the particles of dispersed agent exposed and free to dissolve or otherwise be dispersed into the ambient fluid. Simultaneous solution and subsequent diffusion of the agent by a Higuchi-type mechanism might also accompany this mode of release. Hopfenberg[186] has considered an alternate possibility in which the agent is unable to diffuse through the matrix polymer; however, the environmental fluid is slowly imbibed into the polymer, plasticizes it, and thus the agent becomes free to diffuse from the device. Here, it is the rate of diffusion of the ambient fluid in the polymer which controls the release rather than the rate of diffusion of the agent itself.

Reservoir Devices

An important class of controlled release devices have a reservoir containing the agent to be released which is surrounded by an appropriate polymeric membrane. Clearly, this configuration is more difficult to fabricate than the simpler monolithic or matrix ones; however, it offers some unique opportunities to regulate the rate and pattern of release.

The most common reservoir devices have a continuous membrane coating[168] which is relatively impermeable to water, and release of the agent from the reservoir occurs by its diffusion through this membrane. Clearly, the selection of the membrane material will allow considerable latitude in the rate of release obtained. The membrane might be porous in which case solute permeation occurs through the water contained within these pores. More commonly, the membrane is a fully dense polymer and permeation through it occurs by the usual solution-diffusion mechanism. If all of the agent within the reservoir is dissolved, then the release rate will decline in time since its activity there will decrease, hence the driving force for transport will decrease, as agent is removed. A more interesting possibility is when the agent exists as a suspension within the reservoir, and thus its activity remains constant as the agent is removed so long as saturation is maintained. Since the driving force for transport remains constant, the release rate will be independent of time, i.e., zero order. In this case the flux through the membrane will be given by

$$J = \frac{DC_s}{\ell} \tag{23}$$

where ℓ is the membrane thickness.

Several commercial examples of such devices have been introduced. The ALZA Corporation markets a device called OCUSERT which is placed in the cul de sac of the eye and releases pilocarpine to control glaucoma. Another ALZA product called PROGESTASERT is placed in the uterus and releases progesterone to effect birth control[167]. Each of these are macroscopic devices; however, microencapsulation is a rapidly growing form of the reservoir concept[187–193]. Microcapsules for delivering fertilizers and insecticides are commercially available and many of these employ porous membranes formed by interfacial polymerization[193]. Microcapsules for medical applications have been the object of much research and development activity.

A variation on the reservoir concept is a group of devices which contain on orifice or opening in the membrane. In these, the agent does not diffuse through the membrane, but rather water is osmotically imbibed through the membrane into the reservoir and volumetrically pumps the agent out of the reservoir through the opening and into the environment. For this mode of operation, the membrane should be semi-permeable such that it transports water at the desired rate while being relatively impermeable to the agent within the reservoir. Theeuwes[194] has described a device, called an "elementary osmotic pump", in which the agent is a drug in salt form which provides the osmotic driving force for water transport (see Fig. 12). Such a device

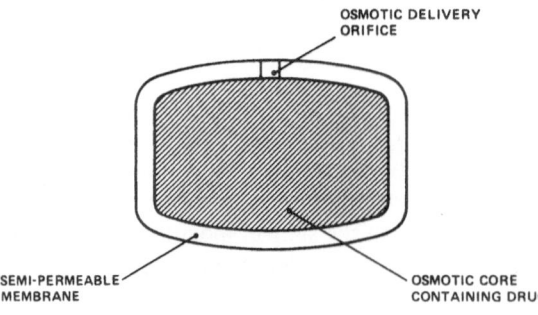

OSMOTIC DELIVERY
ORIFICE

SEMI-PERMEABLE
MEMBRANE

OSMOTIC CORE
CONTAINING DRUG

Fig. 12. Cross-sectional view of elementary osmotic pump (courtesy of J. Pharm. Sci.)

might be introduced orally into the gastrointestinal tract for drug administration. Other variations[167] have been described in which the agent to be delivered does not have to provide the osmotic driving force. Implanation of such devices has been described[167].

Composite Devices

More sophisticated devices have been described which, in effect, are composites of simpler devices designed to offer more control over the rate and pattern of release[168]. A very common example of this is a matrix device containing dispersed agent which is covered by a membrane in which no agent is dispersed. In the usual case, transport through the membrane is considerably slower than would be the Higuchi-type delivery from the matrix itself, and thus the membrane is the rate limiting element. This, in effect, reduces the role of the matrix part to that of a reservoir, but one that is solid, easily formulated, and readily manipulated. A dental device of this type for prolonged administration of fluoride has been described[195].

Chandrasekaran and Shaw[196] have described a highly engineered device for delivery of drugs through the skin. It consists of multiple laminations of Higuchi-type layers and porous membranes designed to give a predetermined temporal pattern of drug release which is not zero order from the device itself but has the desired systemic effect when the permeation characteristics of the skin[197, 198] are considered.

Other Systems

Several delivery systems have been described in which the primary role of the polymer is to serve as a convenient carrier for the agent rather then serve any specific rate controlling function. Ashare et al.[199] have described a hollow fiber system in which the agent is contained within the bore and release is effected by its evaporation and subsequent diffusion through the air above it within the bore. One of its intriguing uses is for release of pheromones, i.e. sex attractants, for control of insects. An alternate concept is release of such agents held within the pore of a suitably constructed open celled polymeric foam[200].

A very interesting approach has been pioneered by Harris et al.[201] in which the agent is chemically attached as pendant groups to a polymer chain which are eventually released to the environment as the connecting bond is broken. A simplified example is the attachment of the insecticide 2,4-D,

to poly(vinyl alcohol) and its subsequent release by hydrolysis. The factors which govern the rate and pattern of release have not been fully identified and may be rather complex. Most studies to date have centered on the complex chemistry involved in synthesizing these materials and some field trials for applications such as weed control.

Removal of Residual Monomer from Polymeric Materials

The ease of migration of small molecules in high polymers has been successfully
exploited for gas separations, water purification and controlled release. These topics
have been discussed earlier in this review. There is also, however a negative side to
the comparative ease of transport in polymers compared with metals and glass. They
are obviously not as good gas and vapor barriers and therefore have definite limita-
tions for certain packaging applications. This subject will not be discussed further in
this presentation.

There is an additional problem however, the migration of small quantities of
residual materials from a polymeric material into its contents. Good examples are
problems involved with the use of polymers for domestic water pipes and for con-
tainers and packaging of various kinds. It has long been a problem that such migra-
tions may bring about unfavorable organoleptic changes in the contents. In addition,
the entry of oxygen could cause spoilage of the contents. These problems have been
largely solved by the correct choice of the polymers and additives. A new and much
more severe problem has, however, arisen in recent years. This is the recognition that
certain components, particularly residual monomers, are highly toxic and in some
cases carcinogenic. Even minute residues of these substances cannot therefore be
tolerated in polymers destined for use in contact with foods or drinking water for
example. Both vinyl chloride and acrylonitrile appear to be clearly in this category
and as polymers in high volume use for plastic bottles and other packaging and for
domestic water pipes they represent a major industrial problem. It is clear that the
residual monomer content of polymers for these applications must be reduced to
vanishingly small amounts.

It is perhaps appropriate at this point to review briefly some of the problems
and technology involved in reducing the residual monomer content of plastic
materials.

The removal of residual monomer from polymers at the final stage of the polymerization process has been standard industrial practice since the early days of
synthetic rubber manufacture. Diene polymerizations are customarily "short stopped"
at comparatively low conversions and the remaining monomer removed by decreasing
the pressure, vacuum stripping or by steam stripping. This is to avoid excessive gelation. With vinyl chloride and vinyl acetate, polymerizations again are often not carried
to completion and the residual monomer is removed in a similar fashion. This has
economic advantages since the polymerization rates are very slow at high conversions
and also the product is improved. In the case of polyvinyl chloride there is greater
particle porosity for example and with polyvinyl acetate there is less branching. In
the case of polystyrene, reduction of the monomer content to, say, less than 0.05%
results in a product with a much higher heat distortion temperature. In this case however two catalysts and/or multi-temperature cycles are practiced to carry the polymerization near completion. It should be repeated that in recent years a new motiva-
tion to reduce the monomer content to vanishingly small amounts has arisen. This is
the knowledge that certain monomers, in particular vinyl chloride and acrylonitrile
have extremely high toxicity and are indeed believed to be carcinogenic. Although
in principle, this reduction of monomer content could be achieved by complete

polymerization, the current approach is to remove the monomer itself by direct or countercurrent steam stripping. With both approaches, the limiting factor becomes the diffusion of the residual monomer in the polymer. The characterization of such transport behavior has become of great importance and some features of this process will be briefly reviewed.

Poly(vinyl chloride), poly(acrylonitrile) and the high acrylonitrile copolymers have presented the major problems with respect to reducing the residual monomer content to extremely low levels. These are in the glassy state under the conditions where monomer removal must be carried out. In principle, the temperature should be raised above the glass temperature to facilitate monomer removal. In practice, however, the systems are usually lattices or slurries of suspension polymer and coagulation could become a problem. In any case, both the rubbery and glassy states must be considered in any discussion of the monomer removal problem. The basic principles of the transport of gases in both situations have been presented briefly and with appropriate literature references in the introductory section of this review. Small organic molecules such as monomers will obey the same laws when present in the very low concentrations which are involved in practice. It is interesting to compare solubilities, for example under one atmosphere of nitrogen, a typical rubber will dissolve about 60 ppm, most polymers dissolve similar amounts. This is actually within the range of residual monomer contents under consideration in most toxic monomer problems.

With rubbery polymers, the transport process is well described by Fick's laws and, at the low solubilities under consideration, with nearly concentration independent diffusion constants. The solubilities also are simple in that Henry's law is obeyed. Diffusion constants are usually in the 10^{-9} to 10^{-11} cm^2/sec range for organic molecules at very low concentrations. With glassy polymers the transport at very low concentration is also Fickian but with the complications introduced by dual mode sorption as discussed in detail in the introductory section. Here, apparent diffusion constants in the range of 10^{-10} to 10^{-14} cm^2/sec are often found.

With problems concerned with the reduction of residual monomer content desorption kinetics are relevant. Fortunately these are normally Fickian but with concentration dependent diffusion constants. The complications of so-called Case II, relaxation controlled diffusion often found in glassy polymers at high concentration gradients[63−79, 202, 203] are also not normally observed in desorption processes.

A convenient measure of desorption rates is the time taken to reduce the concentration to one half of the initial value. Up to the appearance of the monomer residual problem, almost all transport measurements were conducted with flat films. With this geometry the half times, $t_{1/2}$ are given by $t_{1/2} = 0.0492 \dfrac{\ell^2}{D}$ where ℓ is the film thickness in cms and D the diffusion constant in cm^2 per second. Even with a half mil (1.27×10^{-3} cm) film a diffusion constant of 10^{-13} cm^2/sec leads to a half time of more than nine days which made detailed measurements almost impossible.

With residual monomer studies, polymer spheres are mostly involved since the polymers are produced as latices, suspension polymers or powders. The half time with spherical geometry is given by: $0.00766 \dfrac{d^2}{D} = t_{1/2}$ where d is the diameter of the particle.

This reduces the half-time for a one half mil diameter specimen to 34.2 hours still lending a formidable time barrier to detailed studies. However, spherical polymer particles, of reasonable uniform size distribution, are rather readily available down to as low as 0.1 μm diameter. This reduces the half time down to 7.66 seconds and brings transport studies of even the lowest diffusion constant polymers[84] within easy experimental reach. Studies of this kind were pioneered by Berens[82, 204] and later by Enscore et al.[73, 74] with n-hexane in polystyrene. The elegant studies of Berens of the transport of vinyl chloride monomer in polyvinyl chloride have been summarized in a recent review by Berens and Hopfenberg[205] and only the main conclusions will be presented here.

It is clear from the above discussion that the smaller the size of the polymer particle, the faster the residual monomer will diffuse out. This is particularly effective since the rate changes with the square of the particle diameter. This is also well illustrated in Fig. 13. Also included in this figure is the effect of a few (10% by weight) larger particles, it can be seen that these hamper the overall residual monomer reduction to a very large extent. Small, uniform particles, therefore, are essential to ensure efficient monomer removal. The diffusion coefficient for monomer increases strongly with temperature. In the case of VCM in PVC the activation energy for diffusion is 17.0 Kcals per mole. Increasing the temperature from 20 °C to 80 °C for example, increases the diffusion constant by 143 times; this reduces the time for monomer removal by the same factor. Since the small traces of monomer are believed to be largely sorbed in the preexisting microvoids in the glassy polymer and have much lower mobility, it is extremely helpful to raise the temperature above the glass transition temperature, but this approach as mentioned previously, sometimes causes particle agglomeration problems.

Reliable sorption and diffusion data for vinyl chloride monomer in polyvinyl chloride have been reported by Berens[82, 204]. Berens and Hopfenberg[205] have used these values to calculate the rates of release of vinyl chloride monomer from molded

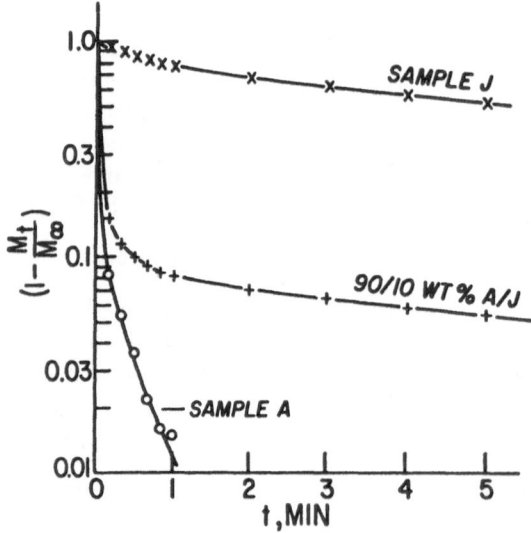

Fig. 13. Experimental late stage desorption curves for VCM diffusion in two different diameter PVC samples and a mixture of the two samples at 90 °C

polyvinyl chloride for four situations which could be encountered under practical conditions, viz:

Case I: Residual vinyl chloride monomer (RVCM) loss during storage of a freshly manufactured PVC product.

Case II: RVCM loss from freshly formed PVC products into a closed, finite medium.

Case III: RVCM loss into a closed, finite medium from previously aged PVC products.

Case IV: RVCM loss from thin-walled bottles into finite closed contents.

Case I is the simplest and can be treated at short times ($M_t/M_\infty < 0.5$) using the following approximate expression for desorption from a slab of initially uniform concentration[176].

$$\frac{M_t}{M_\infty} = \frac{4}{\pi^{1/2}} \left[\frac{\bar{D}t}{\ell^2} \right]^{1/2} \tag{24}$$

where ℓ is the sheet thickness, t is time and $\bar{D}$ is the average effective low concentration diffusion coefficient. M_t is the quantity of monomer desorbed at time t and M_∞ is the total amount of monomer eventually desorbed at infinite time. The numerical results given in Fig. 14 as plots of M (fractional VCM loss = M_t/M_∞) versus $t^{1/2}$ were calculated for several sheet thicknesses at 30 °C for $\bar{D} = 2 \times 10^{-12}$ cm^2/sec using the exact infinite series solution given by Crank[176]. The linear approximation for M vs $t^{1/2}$ given by Eq. (24) fits the data quite well over much of the range of M. The concentration profiles of RVCM at different times for a container with a 1/8" wall at 30 °C

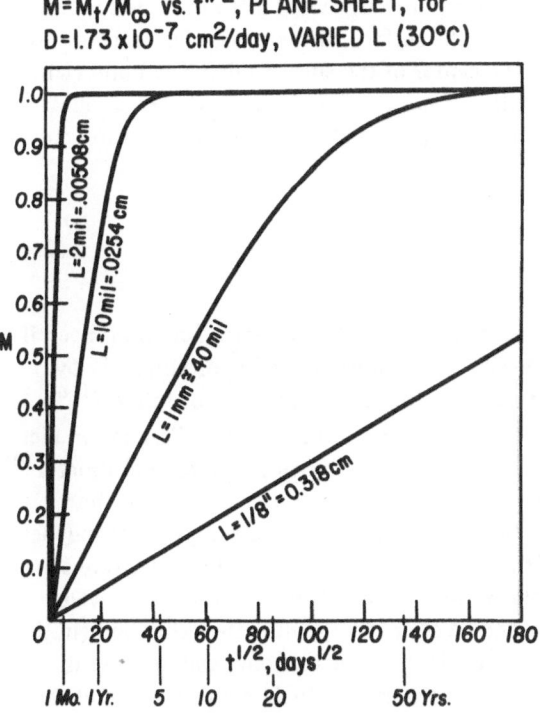

M=M_t/M_∞ vs. $t^{1/2}$, PLANE SHEET, for D=1.73 x 10^{-7} cm^2/day, VARIED L (30°C)

Fig. 14. Theoretical curves of fractional VCM loss (M) versus the square root of time for PVC sheets of various thicknesses (L) for $\bar{D}$ = 2 x 10^{-12} cm^2/sec at 30 °C (Case I)

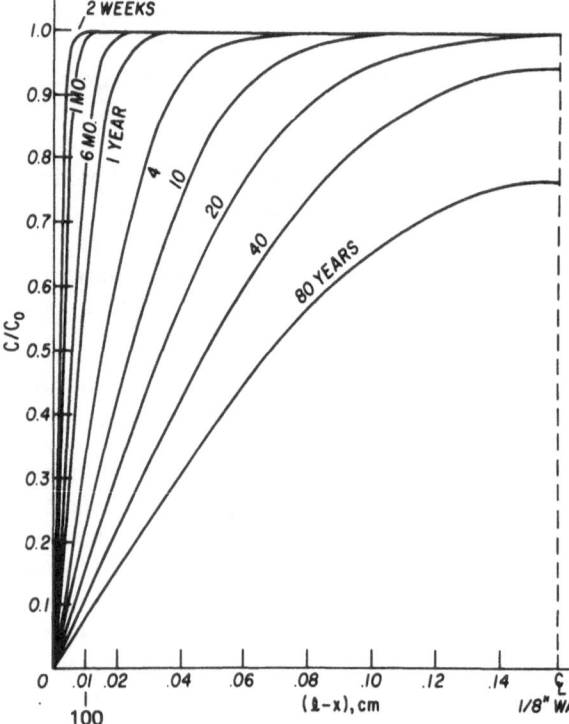

Fig. 15. Relative VCM concentration (C/C_0) vs. depth below the sheet surface ($\ell - x$) at various times, calculated for 1/8 in. thick PVC sheet for $\bar{D} = 2 \times 10^{-12}$ cm^2/sec at 30 °C (Case I)

with $\bar{D} = 2 \times 10^{-12}$ cm^2/sec are shown in Fig. 15. The VCM lost in the first month comes only from the 100 μm of PVC near the surface. It takes over 20 years in this case for the VCM concentration near the center of the sheet to decrease appreciably. For times short enough that the centerline concentration does not decrease significantly below its original values C_0, one can approximate the time and position concentration profile[205] by:

$$\frac{C}{C_0} = \text{erf} \left[\frac{\ell - x}{2 \, (\bar{D}t)^{1/2}} \right] \tag{25}$$

where x is the distance from the slab centerline and erf is the error function. Even if D depends on concentration, eqs. (24) and (25) can be applied approximately by using an average diffusion coefficient over the concentration range of interest[176].

Cases III and IV are of direct interest to the packaging industry. Case IV, in fact, relates directly to monomer migration in blow-molded bottles. Daniels and Procter[206] have treated simultaneous diffusion of VCM into the environment and into the package contents over a time scale sufficient to invalidate the assumption of independent diffusional processes at each of the bottle wall boundaries. Over relatively short time periods, approximately half of the VCM originally sorbed in the bottle wall tends to diffuse into the environment and be swept away while the other half (in the inside half of the wall) tends to diffuse into the package contents and increase its VCM content. Such a simultaneous process increases the VCM concentration at the

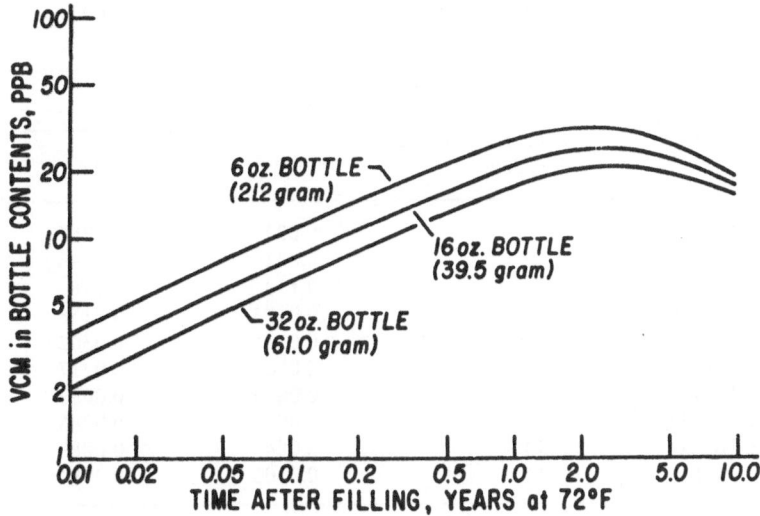

Fig. 16. VCM concentration in water contents of various size PVC bottles initially containing 1 ppm of VCM, uniformly distributed through the bottle wall

inside wall surface and reduces the VCM concentration at the air side wall. Thus, the concentration gradient eventually favors net transport out of the contained medium back into the wall and ultimately into the surrounding air. Typical time response curves describing extraction of VCM from PVC bottles calculated using such a model are presented in Fig. 16. The curves reach maximum values of concentration over time periods which could encompass a reasonable shelf life of the stored contents. Clearly, at infinite time the concentration of VCM in the bottle wall and contents will fall to zero.

The approach discussed above has also been applied by Berens and Daniels[207] to predict the rate of migration of monomer in polyvinyl chloride pipe and has been discussed further by Berens and Hopfenberg[205]. The monomer lost, ΔM, from the polymer in the time interval, a, during a continual open-system storage would be given by

$$\Delta M = 4 \left[\frac{\overline{D}}{\pi L^2} \right]^{1/2} [(t - a)^{1/2} - t^{1/2}] \tag{26}$$

where L is the thickness of the pipe wall and t the age of the polymer before use. Their comparison of calculated results with the experimental results of O'Mara and DeCapita[208, 205] is presented in Table 12. The agreement is quite satisfactory considering the assumptions made and the possible experimental errors. Refs.[205] and [206] give an excellent and expanded treatment of the migration of monomers from glassy polymeric molded or extruded objects.

Recently Koros and Hopfenberg[209] have considered explicitly the effect of dual mode sorption on the local effective concentration dependent diffusion coefficient for low activity penetrant migration in glassy polymers. They showed that

Table 12. Comparison of predicted and experimental extraction results, water-filled 1-in. I.D.,
1/8-in. Wall PVC pipe samples, 23 °C

VCM in pipe mg/kg	Pipe age t	Extraction time a, days	VCM in water, mg/kg	
			Expt.	Calc.
292	~6 Months	3	0.021	0.0257
292	~6 Months	7	0.0414	0.0598
292	~6 Months	14	0.113	0.118
177	~6 Months	3	0.0173	0.0156
177	~6 Months	7	0.0335	0.0362
177	~6 Months	14	0.056	0.0717
22	~6 Months	3	0.0006	0.0019
22	~6 Months	7	0.0022	0.0045
22	~6 Months	14	0.0046	0.0089
29	~1 Year	14	0.0105	0.0084

the average diffusion coefficient for the concentration interval from C_1 to C_2, $\bar{D}$, determined from the slope of M_t/M_∞ plotted according to Eq. (24) is equal to:

$$\bar{D} = \left[\frac{D_D + \dfrac{D_H K}{(1 + \alpha C_{D_2})(1 + \alpha C_{D_1})}}{1 + \dfrac{K}{(1 + \alpha C_{D_2})(1 + \alpha C_{D_1})}} \right] \tag{26}$$

where D_D is the constant diffusion coefficient of the so-called "dissolved" species [C_D in Eq. (7)] and D_H is the constant diffusion coefficient of the Langmuir species, C_H. The parameters $K = C'_H b/k_D$ and $\alpha = b/k_D$ have been discussed earlier. For total desorption experiments, $C_{D_2} \simeq 0$ and the following simplification in $\bar{D}$ results:

$$\bar{D} = \frac{D_D + \dfrac{D_H K}{(1 + \alpha C_{D_1})}}{1 + \dfrac{K}{(1 + \alpha C_{D_1})}} \tag{27}$$

If the initial impurity concentration is also very low, typical of the migration of trace impurities, the expression for $\bar{D}$ can be further simplified to yield a low concentration limiting value, $\bar{D}_0$, which is a true constant:

$$\bar{D}_0 = \frac{D_D + D_H K}{1 + K} = \frac{D_D(1 + FK)}{1 + K} \tag{28}$$

where $F \equiv D_H/D_D$ measures the inherent mobility ratio of trace impurity in the two sorption modes.

A value of D_D (which is by definition concentration independent) can be calculated if a value of F is specified or known for systems conforming to dual mode sorption using measured values of $\bar{D}$, K and α. Berens[82] has demonstrated that a dual mode sorption isotherm describes the low concentration region of VCM/PVC sorption. The Berens study provides a value of K = 5.41 and $\alpha = 1.5 \times 10^{-3}$ (ppm^{-1}). Since a value of F has not been measured for this system, the migration kinetics for both limits (F = 1 and F = 0) were calculated by Koros and Hopfenberg for equilibrium parameters K = 5.41 and $\alpha = 1.5 \times 10^{-3}$ ppm^{-1}. These results are presented graphically in Fig. 17. The values of D_D were calculated from the measured value of $\bar{D} = 2 \times 10^{-12}$ cm^2/sec for soprtion of VCM in PVC over the concentration interval 0–2000 ppm at 30 °C[82]. The $\bar{D}$ was used in Eq. (27) with the assumed value of F = 0 or 1.0 along with the aforementioned values of the equilibrium parameters reported by Berens[204]. The zero concentration limit, of $\bar{D}$ was subsequently calculated from Eq. (28) and used in the calculations to prepare Fig. 17. Although total immobilization of Langmuirian species significantly retards migration, complete evacuation of penetrant from the polymer is achieved for both limiting cases at sufficiently long times if the two populations, C_D and C_H are in local equilibrium.

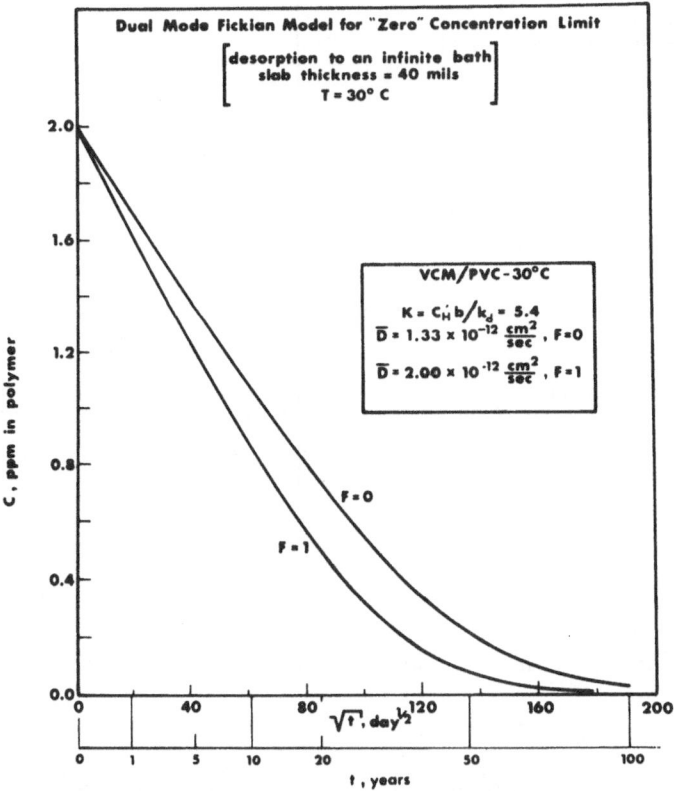

Fig. 17. Calculated effect of degree of "mobilization",
F = D_H/D_D on the release kinetics of VCM from a PVC slab at 30 °C[209]

If the Langmuirian species are *not* in local equilibrium with the Henry's law species an additional retardation will result due to the kinetic limitation preventing equilibrium between Henry's law and Langmuirian molecules. Completly irreversible sorption would lead to "trapping": however, any reasonable non-zero rate of transfer between holes and matrix would be achieved within some academically long time scale of desorption. These notions have been treated by Tshudy and von Frankenberg[210] for the case of F = 0 and discussed qualitatively by Koros and Hopfenberg[209] for cases where $0 < F < 1$.

The effect of the Langmuirian sorption mode *per se* on migration kinetics is explored further in Fig. 18. The total immobilization limit (F = 0) has been arbitrarily selected in preparing curve (a) of Fig. 18. Clearly, other values of F could be treated as well. The K = 0 limit (Henry's law sorption) gives rise to much more rapid desorption than for K > 0 since the mean diffusion coefficient corresponding to Eq. (27) is equal to D_D when K = 0. For all values of K > 0, $\bar{D} < D_D$ and desorption is retarded, however total evacuation of the sample is eventually predicted at times which, although non-infinite, are indeed extraordinarily long.

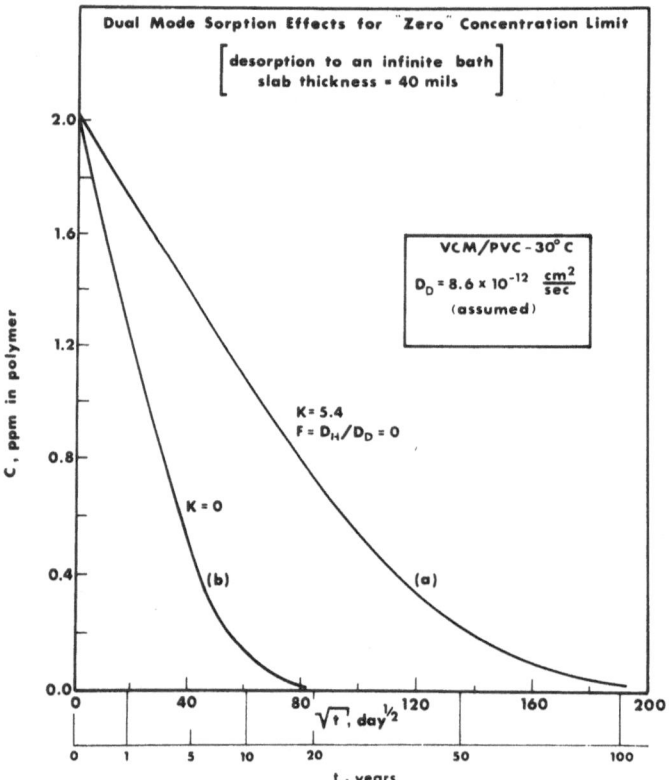

Fig. 18. Comparison of desorption kinetics for the case where only Henry's law sorption occurs (K = 0 as in a rubbery polymer) with the case in which Langmuirian and Henry's law sorption occur (K > 0) in local equilibrium with each other for the case of F = 0 [209]

References

1. Graham, T.: Phil. Mag. *32*, 401 (1866)
2. Meares, P.: J. Am. Chem. Soc. *76*, 3415 (1954)
3. Von Wroblewski, S.: Wiedemanns Ann. Physik. *8*, 29 (1879)
4. Barrer, R. M.: Trans. Faraday Soc. *35*, 628 (1939)
5. Kumins, C. A., Kwei, T. K.: Diffusion in polymers, Chap. 4. Crank, J., and Park, G. S. (eds.). New York: Academic Press 1968
6. Michaels, A. S., Bixler, H. J.: J. Poly. Sci. *50*, 393 (1961)
7. Van Amerongen, G. J.: Rubber Chem. Tech. *37*, 1065 (1964)
8. Stannett, V.: Ref. 5, Chap. 2
9. Stannett, V.: J. Membrane Sci. *3*, 97 (1978)
10. Meares, P.: Trans. Faraday Soc. *53*, 101 (1957)
11. Barrer, R. M., Chio, H. T.: J. Poly. Sci. *C10*, 111 (1965)
12. Lundstrom, J. E., Bearman, R. J.: J. Poly. Sci. Physics Ed. *12*, 927 (1974)
13. Fujii, M., Hopfenberg, H. B., Stannett, V.: J. Macromol. Sci. Physics Ed. *B15(3)*, 421 (1978)
14. Burgess, W. H., Hopfenberg, H. B., Stannett, V.: Macromol. Sci. Physics Ed. B5(1), 23 (1971)
15. Allen, S. G.: Ph. D. Thesis, North Carolina State University, 1975
16. Daynes, H. A.: Proc. Roy. Soc. (London) A. *97*, 286 (1920)
17. Van Amerongen, G. J.: J. Appl. Phys. *17*, 972 (19 46)
18. Frisch, H. L.: J. Phys. Chem. *61*, 93 (1957)
19. Frisch, H. L.: J. Phys. Chem. *62*, 401 (1958)
20. Frisch, H. L.: J. Chem. Phys. *37*, 2408 (1962)
21. Petroupolos, J. H., Rousis, P. P.: J. Polym. Sci., Part C *22*, 917 (1969)
22. Barrer, R. M., Skirrow, G.: J. Poly Sci. *3*, 564 (1948)
23. Bixler, H. J., Sweeting, O. J. in: The science and technology of polymer films, Vol. II. Sweeting, O. J. (ed.). New York: John Wiley and Sons 1971
24. Stern, S. A., Mulhaupt, J. T., Gareis, P. J.: A. I. Ch. E. J. *15*, 64 (1969)
25. Suwandi, M. S., Stern, S. A.: J. Poly. Sci. Physics Ed. *11*, 663 (1973)
26. Stiel, L. I., Harnish, D. F.: A. I. Ch. E. J. *22*, 117 (1976)
27. See for example Reid, R. C., Prausnitz, J. M., Sherwood, T. K.: The properties of gases and liquids, p. 12. New York: McGraw Hill 1977
28. See for example the various chapters in Ref.[5]
29. Matthes, A.: Kolloid Z. *108*, 79 (1944)
30. Barrer, R. M., Barrie, J. A., Slater, J.: J. Poly. Sci. *27*, 177 (1958)
31. Chan, A. H., Paul, D. R., Koros, W. J.: J. Membrane Sci. *3*, 117 (1978)
32. Michaels, A. S., Vieth, W. R., Barrie, J. A.: J. Appl. Phys. *34*, 1, 13 (1963)
33. Vieth, W. R., Sladek, K. J.: J. Coll. Sci. *20*, 1014 (1965)
34. Vieth, W. R., Howell, J. M., Hseih, J. H.: J. Membrane Sci. *1*, 177 (1976)
35. Paul, D. R.: J. Poly. Sci. *A2(7)*, 1811 (1969)
36. Paul. D. R., Kemp, D. R.: J. Poly. Sci. *C41*, 79 (1973)
37. Petropoulos, J. H.: J. Poly. Sci. *A2(8)*, 1797 (1970)
38. Paul, D. R., Koros, W. J.: J. Poly. Sci. – Phys. Ed. *14*, 675 (1976)
39. Koros, W. J., Chan, A. H., Paul, D. R.: J. Membrane Sci. 2, 165 (1977)
40. Koros, W. J., Paul, D. R., Rocha, A. A.: J. Poly. Sci. Phys. Ed. *14*, 687 (1976)
41. Koros, W. J., Paul, D. R.: Polym. Engr. and Sci., in press
42. Koros, W. J., Paul, D. R.: J. Polym. Sci., Polym. Phys. *16*, 2171 (1978)
43. Assink, R. A.: J. Poly. Sci.-Phys. Ed. *13*, 1665 (1975)
44. Stannett, V., Williams, J. L.: J. Poly. Sci. *C10*, 45 (1966)
45. Kumins, C. A., Roteman, J.: J. Poly. Sci. *55*, 683 (1961)
46. Tikhomirov, B. P., Hopfenberg, H. B., Stannett, V., Williams, J. L.: Makromol. Chem. *118*, 117 (1968)
47. Ziegl, K. D., Eirich, F. R.: J. Poly. Sci. Phys. Ed. *12*, 1127 (1974)

48. Ziegl, K. D., Frensdorff, H. K., Blair, D. E.: J. Poly. Sci. A2(7), 809 (1969)
49. Enns, J. B., Simha, R.: J. Macromol. Sci. Physics *B13*, 11 (1977)
50. Boyer, R. F.: J. Macromol. Sci. Physics *B8*, 503 (1973)
51. Boyer, R. F.: J. Poly. Sci. *C50*, 189 (1975)
52. Yasuda, H., Stannett, V.: J. Macromol. Sci. Physics *B3*, 589 (1969)
53. Armstrong, A. A., Wellons, J. D., Stannett, V.: Makromol. Chemie. *95*, 78 (1966)
54. Barrie, J. A.: Chap. 8 in Ref.[5]
55. Vieth, W. R., Douglas, A. S., Bloch, R.: J. Macromol. Sci. Physics *B3*, 737 (1969)
56. Stannett, V., Haider, M. I., Koros, W. J., Hopfenberg, H. B.: Polymer Sci. and Engr. in
    press
57. Starkweather, H. W.: J. Appl. Poly. Sci. *2*, 129 (1959)
58. Yasuda, H., Stannett, V.: J. Poly. Sci. *57*, 907 (1962)
59. Wellons, J. D., Stannett, V.: J. Poly. Sci. A1(4), 593 (1966)
60. Williams, J. L., Hopfenberg, H. B., Stannett, V.: J. Macromol. Sci. Physics *B3*, 711 (1969)
61. Orofino, T. A., Hopfenberg, H. B., Stannett, V.: J. Macromol. Sci. Physics *B3*, 777 (1969)
62. Starkweather, H. W. in: Structure-solubility relationships in polymers. Harris, F. W. and
    Seymour, R. B. (eds.). New York: Academic Press, Inc. 1977
63. Alfrey, T.: Chem. Engng. News, *43*, 64 (1965)
64. Alfrey, T., Gurnee, E. F., Lloyd, W.: J. Polymer Sci. C12, 249 (1966)
65. Michaels, A. S., Bixler, H. J., Hopfenberg, H. B.: J. Appl. Polymer Sci. *12*, 991 (1968)
66. Hopfenberg, H. B., Holley, R. H., Stannett, V.: Polymer Eng. Sci. *9*, 242 (1969)
67. Bray, J., Hopfenberg, H. B., Stannett, V.: Polymer Eng. Sci. *10*, 376 (1970)
68. Baird, B. R., Hopfenberg, H. B., Stannett, V.: Polymer Eng. Sci. *11*, 274 (1971)
69. Hopfenberg, H. B., Stannett, V., Folk, G. N.: Polymer Eng. Sci. *15*, 261 (1975)
70. Jacques, C. H. M., Hopfenberg, H. B., Stannett, V.: Polymer Eng. Sci., *13*, 81 (1973)
71. Jacques, C. H. M., Hopfenberg, H. B.: Polymer Eng. Sci. *14*, 441 (1974)
72. Jacques, C. H. M., Hopfenberg, H. B.: Polymer Eng. Sci. *14*, 449 (1974)
73. Enscore, D. J., Hopfenberg, H. B., Stannett, V.: Polymer *18*, 793 (1977)
74. Enscore, D. J., Hopfenberg, H. B., Stannett, V.: Polymer *18*, 1105 (1977)
75. Nicolais, L., Drioli, E., Hopfenberg, H. B., Tidone, D.: Polymer *18*, 1137 (1977)
76. Nicolais, L., Drioli, E., Hopfenberg, H. B., Caricati, G.: J. Membr. Sci. *3*, 231 (1978)
77. Thomas, N., Windle, A. H.: Polymer *19*, 255 (1978)
78. Thomas, N., Windle, A. H.: Polymer *18*, 1195 (1977)
79. Thomas, N., Windle, A. H.: J. Membr. Sci. *3*, 337 (1978)
80. Sarti, G. C.: Solvent Osmotic Stresses and the Prediction of Case II Transport Kinetics,
    Polymer, in press
81. Berens, A. R.: Polymer *18*, 697 (1977)
82. Berens, A. R.: Angew. Makromol. Chem. *47*, 97 (1975)
83. Berens, A. R.: J. Macromol. Sci. (B), in press
84. Berens, A. R.: J. Membr. Sci. *3*, 247 (1978)
85. Berens, A. R., Hopfenberg, H. B.: Polymer *19*, 489 (1978)
86. Fujita, H.: Ref.[5], p. 75
87. Park, G. S.: Ref.[5], p. 141
88. Frisch, H. L.: Polymer Engr. Sci., in press
89. Hopfenberg, H. B., Paul, D. R. in: Polymer blends, Vol. 1, p. 445. Paul, D. R. and New-
    man, S. (eds.). New York: Academic Press 1978
90. Kammermeyer, K. in: Progress in separation and purification, Vol. I, p. 335. Perry, E. S.
    (ed.). New York: Interscience 1968
91. Li, N. N., Long, R. B., Henley, E. J.: Ind. Engng. Chem. *57(3)*, 18–29 (1965)
92. Michaels, A. S., Bixler, H. J. in: Progress in separation and purification, Vol. I., p. 134.
    Perry, E. S. (ed.). New York: Interscience 1968
93. Hwang, S., Kammermeyer, K.: Membranes in separations. New York: John Wiley and Sons
    1975
94. Stern, S. A. in: Membrane separation processes, Chap. 8. Meares, P. (ed.). New York:
    Elsevier 1976

95. Rogers, C. E., Fels, M., Li, N. N. in: Recent developments in separation science, Vol. III, p. 107. Li, N. N. (ed.). Ohio: C. R. C. Press, Cleveland 1972
96. DuPont, Permasep Technical Bulletin Nos. 105 and 110, 1972
97. McCandless, F. P.: Ind. Engng. Chem., Proc. Des. Dev. *11*, 470 (1972)
98. Stern, S. A., Walawender, W.: Separation Sci. *4*, 129 (1969)
99. "General electric permselective membranes", published by Medical Development Operation Chemical and Medical Division, General Electric Co.
100. Stern, S. A., Sinclair, T. F., Gareis, P. J., Vahldieck, N. P., Mohr, P. H.: Ind. Engr. Chem. 57, 49 (1965)
101. Hwang, S. T., Choi, C. K., Kammermeyer, K.: Separation Science *9*, 461 (1974)
102. Pye, D. G., Hoehn, H. H., Panar, M.: J. Appl. Poly. Sci. *20*, 1921 (1976)
103. Paul, D. R.: I. & E. C. Proc. Des. and Dev. 10, 375 (1971)
104. Ash, R., Barrer, R. M., Sharma, P.: J. Membr. Sci. *1*, 17 (1976)
105. Pye, D. G., Hoehn, H. H., Panar, M.: J. Appl. Poly. Sci. *20*, 287 (1970)
106. Pilato, L., Litz, L., Hargitay, B., Osborne, R. C., Farnham, A., Kawakami, J., Fritze, P., McGrath, J.: A. C. S. Preprints *16(2)*, 42 (1975)
107. Loeb, S., Sourirajan, S.: Univ. Calif. at Los Angeles, Engin. Report No. 60–60 (1961)
108. Loeb, S., Sourirajan, S.: Adv. Chem. Ser., *38*, 117 (1963)
109. Gittens, G. J., Hitchcock, P. A., Sammon, D. C., Wakley, G. E.: Desalination, *8*, 369 (1970)
110. Panar, M., Hoehn, H., Hebert, R.: Macromolecules *6*, 777 (1973)
111. Alegranti, C. W., Pye, D. G., Hoehn, H. H., Panar, M.: J. Appl. Polym. Sci. *19*, 1475 (1975)
112. Blais, P. in: Reverse osmosis and synthetic membranes, Chapt. 9. Sourirajan, S. (ed.). Ottawa, Canada: National Research Council Canada Publications 1977
113. Kimura, S., Sourirajan, S.: A. I. Ch. E. J. 13, 497 (1967)
114. Lonsdale, H. K., Merten, U., Riley, R. L.: J. Appl. Polym. Sci. *9*, 1341 (1965)
115. Gardner, R. J., Crane, R. A., Hannan, J. F.: Chem. Engr. Progress *73*, 78 (1977)
116. Antonson, C. R., Gardner, R. U., King, C. F., Ko, D. Y.: Ind. Eng. Chem., Proc. Des. Dev. *16*, 463 (1977)
117. Zsigmondy, R., Bachmann, W.: Z. Anorg. Allgem. Chem. *103*, 119 (1918)
118. Goetz, A.: U. S. Patent No. 2, 926, 104, 1960
119. Kesting, R. E.: Synthetic polymeric membranes. New York: McGraw-Hill Book Co. 1971
120. Strathmann, H., Kock, K.: Desalination *21*, 241 (1977)
121. Kremen, S. S., Ref.[112], pp. 371–385
122. Baum, B., Holley, Jr., W., White, R. A.: Ref.[94], pp. 187–228
123. Richter, J. W., Hoehn, H. H., U. S. Patent No. 3, 567, 632, 1971
124. Frommer, M. A., Murday, J. S., Messalem, R. M.: European Poly. J. *9*, 367 (1973)
125. Cadotte, J. E., Rozelle, L. T., Petersen, R. J., Francis, P. S., in: Turbak, A. F. (ed.), Membranes from Cellulose and Cellulose Derivatives, Applied Polymer Symposium No. 13, Interscience Publishers, New York, 1970, pp. 73–83
126. Model, F. S., Lee, L. A., in: Reverse osmosis membrane research, pp. 285–297. Lonsdale, H. K., Podall, H. E. (eds.). New York: Plenum Press 1972
127. Goldsmith, R. L., Wechsler, B. A., Hara, S., Mori, K., Taketani, Y.: Desalination *22*, 311 (1977)
128. Strathmann, H., Michaels, A. S.: Desalination *21*, 195 (1977)
129. Walch, A., Lukas, H., Klimmek, A., Pusch, W.: J. Poly. Sci. *12*, 697 (1974)
130. McKinney, Jr., R., Rhodes, J. H.: Macromolecules *4*, 633 (1971)
131. Walch, A., Klimmek, A., Pusch, W.: J. Poly. Sci. *13*, 701 (1975)
132. King, W. M., Hoernschemeyer, D. L., Saltonstall, Jr., C. W. in: Reverse osmosis membrane research, pp. 131–161. Lonsdale, H. K., Podall, H. E. (eds.). New York: Plenum Press 1972
133. Riley, R. L., Lonsdale, H. K., Lyons, C. R., Merten, U.: J. Applied Poly. Sci. *11*, 2143 (1967)
134. Riley, R. L., Fox, R. L., Lyons, C. R., Milstead, C. E., Seroy, M. W., Tagami, M.: Desalination *19*, 113 (1976)

135. Rozelle, L. T., Cadotte, J. E., Cobian, K. E., Kopp, Jr., C. V. in: Reverse osmosis and synthetic membranes, pp. 249–261. Sourirajan, S. (ed.). Ottawa: National Research Council Canada 1977

136. Cadotte, J. E., Cobian, K. E., Forester, R. H., Petersen, R. J.: *In situ*-formed condensation polymers for reverse osmosis membranes. Minneapolis, MN: North Star Research Division 1975

137. Johnson, J. S., Kraus, K. A., Fleming, S. M., Cochran, Jr., H. D., Perona, J. J.: Desalination *5*, 359 (1968)

138. Yasuda, H., Lamaze, C. E.: J. Applied Poly. Sci. *17*, 201 (1973)

139. Merten, U.: Desalination *1*, 297 (1966)

140. Gardner, C. R., Weinstein, J. N., Caplan, S. R.: Desalination *12*, 19 (1973)

141. Leitz, F. B.: Desalination *13*, 373 (1973)

142. Platt, K. L., Schindler, A.: Die Angewandte Makromolekulare Chemie *19*, 135 (1971)

143. Baker, R. W., Strathmann, H.: J. Applied Poly. Sci. *14*, 1197 (1970)

144. Goldsmith, R. L.: Ind. Eng. Chem. Fundam. *10*, 113 (1971)

145. Witmer, F. E.: Environmental Sci. and Technology *7*, 314 (1973)

146. Goldsmith, R. L., deFilippi, R. P., Hossain, S., Timmins, R. S. in: Membrane processes in industry and biomedicine, pp. 267–300. Bier, M. (ed.). New York: Plenum Press 1971

147. Bailey, P. A.: Filtration and Separation *14*, 213 (1977)

148. Bierenbaum, H. S., Isaacson, R. B., Druin, M. L., Plovan, S. G.: I & EC Product Research and Development *13*, 2 (1974)

149. Fleischer, R. L., Alter, H. W., Furman, S. C., Price, P. B., Walker, R. M.: Science *178*, 255 (1972)

150. Frankenfeld, J. W., Li, N. N. in: Recent developments in separation science, Vol. III-B, pp. 285–292. Li, N. N. (ed.). Cleveland, Ohio: CRC Press, Inc. 1977

151. Baker, R. W., Tuttle, M. E., Kelly, D. J., Lonsdale, H. K.: J. Memb. Sci. *2*, 213 (1977)

152. Caracciolo, F., Cussler, E. L., Evans, D. F.: AIChE J. *21*, 160 (1975)

153. Largman, T., Sifniades, S.: Hydrometallurgy *3*, 153 (1978)

154. Tanquary, A. C., Lacey, R. E. (eds.): Controlled release of biologically active agents. Vol. 47 of Adv. Experimental Medicine and Biology. New York: Plenum Press 1974

155. Cardarelli, N. F. (ed.): Controlled release pesticide symposium, University of Akron, 1974

156. Harris, F. W. (ed.): Proc. 1975 Internat. Controlled Release Pesticide Symposium, Wright State University, Dayton, Ohio, 1975

157. Williams, A.: Sustained release pharmaceuticals. Park Ridge, N. J. Noyes Development Corp. 1969

158. Allan, G. G., Chopra, C. S., Friedhoff, J. F., Gara, R. I., Maggi, M. W., Neogi, A. N., Roberts, S. C., Wilkens, R. M.: Chemtech *3*, 171 (1973)

159. Colbert, J. C.: Controlled action drug forms. Park Ridge, N. J., Noyes Data Corp., 1974

160. Cardarelli, N. F.: Chemtech *5*, 482 (1975)

161. Baker, R. W., Lonsdale, H. K.: Chemtech *5*, 668 (1975)

162. Cardarelli, N. F.: Controlled release pesticides formulations. Cleveland, Ohio: CRC Press 1976

163. Paul, D. R., Harris, F. W. (eds.): Controlled Release Polymeric Formulations (ACS Symposium Series Vol. 33), Amer. Chem. Soc., Washington, 1976

164. Flynn, G. L., Yalkowski, S. H., Roseman, T. J.: J. Pharmac. Sci. *63*, 479 (1974)

165. Chemburkar, P. B.: Evaluation, Control, and Prediction of Drug Diffusion Through Polymeric Membranes, Ph. D. Dissertation, University of Florida, 1967

166. Neogi, S. A. N.: Polymer Selection for Controlled Release Pesticides, Ph. D. Dissertation, University of Washington, 1970

167. Michaels, A. S.: Therapeutic systems for controlled administration of drugs: A new application of membrane science, pp. 409–421 in Permeability of plastic films and coatings to gases, vapors and liquids. Hopfenberg, H. B. (ed.), Vol. 6 of Polymer science and technology series. New York: Plenum Press 1974

168. Paul, D. R.: ACS Symp. Ser. *33*, 1 (1976)

169. Cardarelli, N. F.: ACS Symp. Ser. *33*, 208 (1976)

170. Janes, G. E.: ACS Symp. Ser. *33*, 231 (1976)
171. Castelli, V. J., Yeager, W. L.: ACS Symp. Ser. *33*, 239 (1976)
172. Leeper, H., Benson, H.: Polym. Engr. Sci. *17*, 42 (1977)
173. Barrer, R. M.: Diffusion in and through solids. London: Cambridge Univ. Press 1941
174. Jost, W.: Diffusion in solids, liquids and gases. New York: Academic Press 1960
175. Crank, J., Park, G. S. (eds.): Diffusion in polymers. New York: Academic Press 1968
176. Crank, J.: The mathematics of diffusion, 2nd ed. Oxford: Clarendon Press 1975
177. Higuchi, T.: J. Pharm. Sci. *50*, 874 (1961)
178. Higuchi, T.: J. Pharm. Sci. *52*, 1145 (1963)
179. Paul, D. R., McSpadden, S. K.: J. Membrane Sci. *1*, 33 (1976)
180. Roseman, T. J., Higuchi, W. I.: J. Pharm. Sci. *59*, 353 (1970)
181. Chien, Y. W., Lambert, H. J., Grant, D. E.: J. Pharm. Sci. *63*, 365, 515 (1974)
182. Cobby, J., Mayersohn, M., Walker, G. C.: J. Pharm. Sci. *63*, 725 (1974)
183. Brooke, D., Washkuhn, R. J.: J. Pharm. Sci. *66*, 159 (1977)
184. Langer, R., Folkman, J.: ACS Polymer Preprints *18(2)*, 379 (1977)
185. McGinity, J. W., Combs, A. B., Mehta, C. S.: Formulation Factors Influencing the *In Vitro* Release of Morphine Sulfate from Silastic Pellets, Proc. First Internat. Conf. Pharmac. Technology, Vol. 2, D. Duchene, Editor, University of Paris – SUD, Paris, France, 1977, pp. 235–242
186. Hopfenberg, H. B.: ACS Symp. Ser. *33*, 26 (1976)
187. Chang, T. M. S.: Artificial Cells, C. C. Thomas, Springfield, Ill., 1972
188. Gutcho, M.: Capsule technology and microencapsulation. Park Ridge, N. J. Noyes Data Corp. 1972
189. Vandegaer, J. E. (ed.): Microencapsulation – process and applications. New York: Plenum Press 1974
190. Fanger, G. O.: Chemtech *4*, 397 (1974)
191. Goodwin, J. T., Somervill , G. R.: Chemtech *4*, 623 (1974)
192. Chang, T. M. S.: Chemtech *5*, 80 (1975)
193. Thies, C.: Polymer-Plast. Technol. Eng. *5(1)*, 1 (1975)
194. Theeuwes, F.: J. Pharm. Sci. *64*, 1987 (1975)
195. Halpern, B. D., Solomon, O., Kopec, L., Korostoff, E., Ackerman, J. L.: ACS Symp. Ser. *33*, 135 (1976)
196. Chandrasekaran, S. K., Shaw, J. E.: Design of transdermal therapeutic systems, in: Contemporary topics in polymer science, Vol. 2. Pearce, E. M., Schaefgen, J. R. (eds.). New York: Plenum Press 1977
197. Michaels, A. S., Chandrasekaran, S. K., Shaw, J. E.: AIChE J. *21*, 985 (1975)
198. Idson, B.: J. Pharm. Sci. *64*, 901 (1975)
199. Ashare, E., Brooks, T. W., Swenson, D. W.: ACS Symp. Ser. *33*, 273 (1976)
200. Obermeyer, A. S., Nichols, L. D.: ACS Symp. Ser. *33*, 303 (1976)
201. Harris, F. W., Aulabaugh, A. E., Case, R. D., Dykes, M. K., Feld, W. A.: ACS Symp. Ser. *33*, 222 (1976)
202. Hopfenberg, H. B., Stannett, V.: Chap. 9 in: The physics of glassy polymers. Haward, R. N. (ed.). London: Applied Science Publishers 1973
203. Stannett, V., Hopfenberg, H. B., Petropoulos, J. H.: Chap. 8 in: Macromolecular science. Bawn, C. E. H. (ed.). London: Butterworths Press 1972
204. Berens, A. R.: Polym. Engr. and Sci., in press
205. Berens, A. R., Hopfenberg, H. B. in: Recent developments in separation science, pp. 293–312. Norman, Li (ed.). Cleveland, Ohio: C. R. C. Press 1978
206. Daniels, G. A., Proctor, D. E.: Mod. Packaging, April 1975, pp. 45–48
207. Berens, A. R., Daniels, C. A.: Polym. Sci. Eng. *16*, 552 (1976)
208. O'Mara, M. M., DeCapita, E. G.: Internal rep., B. F. Goodrich Company, 1975
209. Koros, W. J., Hopfenberg, H. B.: I. & E. C. Prod. R. & D., (submitted for publication)
210. Tshudy, J. A., von Frankenberg, C.: J. Poly. Sci. Physics Ed. *11*, 2027 (1973)

Molecular Interactions and Macroscopic Properties of Polyacrylonitrile and Model Substances

Gisela Henrici-Olivé and Salvador Olivé

Monsanto Triangle Park Development Center Inc., Research Triangle Park, North Carolina 27709, U. S. A.

The relationships between intra- and intermolecular forces in polyacrylonitrile on the one hand, and the macroscopic behavior of the polymer and fibers thereof on the other, are reviewed. Characteristic properties such as the very high polymerization rate constant in water, the dissolution of the polymer in concentrated inorganic salt solutions, the high melting point, the strong depression of melting point and glass transition temperature by water, the plasticization by polar additives, etc., are traced back to their molecular origins, in particular to the strong intra- and intermolecular noncovalent bonding caused by the highly polar nitrile group. The effects of dipole-dipole interaction, hydrogen bonding and electron-donor-acceptor complex formation are discussed separately.

Table of Contents

1 Introduction

Acrylic fibers have a market share of about 20% in the tremendous worldwide production of synthetic fibers, which presently amounts to some 7×10^6 metric tons per year[1]. The acrylic fibers are made from polymers composed of at least 85% by weight of acrylonitrile, modified by one or more other monomers. Despite the large volume production of acrylic fibers, much of the polymer chemistry and fiber physics involved, and particularly of the inter-relations between both, is still far from being fully understood.

In the present article an attempt is made to collect, and interpret on a molecular basis, what is known thus far about the interactions of polyacrylonitrile molecules with each other and with other molecules, and to relate such interactions to macroscopic properties of polymer and fiber.

The dominant characteristic of the PAN molecule is the presence of the strongly polar nitrile groups, at an intramolecular distance of only a few tenths of a nm. The CN groups have a variety of possibilities to interact with their surroundings. The high dipole moment (3.9 Debye[2]) causes strong attraction or repulsion (according to the orientation) of other molecules, or substituents on molecules, possessing also a high dipole moment. Moreover, the lone pair orbital situated at the nitrogen and oriented 180 ° away from the C≡N bond, is perfectly suited to participate in hydrogen bonding with water or other Brønstedt acids, as well as in electron-donor-acceptor complex formation with Lewis acids. And finally electrons in the π-orbitals of the C≡N triple bond are available for ethylene- or acetylene-like interactions, e. g., with transition metal ions. We shall discuss these effects separately. Frequently we shall have recourse to model substances of low molecular weight (e. g. acetonitrile) for introducing the relevant parameters before we try to interpret the behavior of the polymer or the fiber.

2 Dipole-Dipole Interaction of Nitrile Groups

2.1 Low Molecular Weight Models

Certain physical properties of organic nitriles, such as high boiling point and viscosity[3], temperature dependence of the dielectric polarization[4] and of vibrational spectra[5], have been traced back to aggregation of the nitrile molecules by interaction of their C≡N groups. A strong and rather specific interaction between pairs of CN groups has been suggested by Saum[3]. Experimental data[4, 6] and theoretical calculations[7, 8] indicate an antiparallel alignment of the C≡N groups of interacting pairs of molecules:

R–C≡N
⋮ ⋮
N≡C–R

Equilibria between monomeric and dimeric species are generally assumed, although some higher aggregation cannot completely be ruled out[8]. Taking into account only the monomer-dimer equilibrium:

$$2 \text{ R-CN} \rightleftarrows (\text{R-CN})_2 - E_D$$

the energy of interaction E_D has been estimated. Some data for acetonitrile, which may be taken in first approximation as a model for PAN, are given in Table 1. It also has been estimated that in pure acetonitrile some 70%[5] to 90%[3] of the compound is in the associated state.

Table 1. Energy of dimerization, E_D, of acetonitrile

Experimental or calculation basis	$-E_D$ (kJ mol^{-1})	Ref.
Infrared spectroscopy	22	5)
Dielectric polarization (gase phase)	16	4)
Second virial coefficient (gas phase)	19	9)
Second virial coefficient (gas phase)	22	10)
Viscosity and heat of vaporization	38	3)
CNDO/2 (calculated)	19	7)

Table 2. Stretching frequency of the C≡N bond for some nitriles[11]

Nitrile	$\nu_{C\equiv N}(cm^{-1})$		
	Vapor	Liquid	Solution[a]
CH_3CN	2267	2252	2255
CH_3CH_2CN	2265	2247	2249
$(CH_3)_2CHCN$	2264	2245	2247

[a] In CCl_4.

According to molecular orbital calculations, the interaction is mainly electrostatic in character. Only minor variations of the gross atomic population in the associate, as compared with the isolated molecule, are found[7, 8]. Consistent with this view, the experimental stretching frequency of the CN bond of organic nitriles in the vapor phase (where supposedly the monomeric form predominates) and in liquid state (where 70–90% are associated) are very similar, as shown in Table 2. If the small decrease of 15–20 cm^{-1} is taken as an indication for dimer formation, it may be concluded that the degree of association is similar in the pure liquid and in CCl_4 solution.

2.2 Intra- and Intermolecular Interactions of CN in Polyacrylonitrile

According to present knowledge, the forces effective in polyacrylonitrile, in the unstrained state as well as under stress, result predominantly from interactions of the strongly polar CN groups[12, 13]. Adjacent nitrile groups of the same macromolecule repel each other. This intramolecular repulsion compels the individual macromolecules into a somewhat irregular helical conformation[14]. As indicated in Fig. 1, the twisted, kinked molecule may be thought of as a more of less rigid structure, fitting within a cylinder of about 600 pm diameter. Some of the CN groups will extend beyond the confines of the cylinder. These groups are potentially available for intermolecular dipole-dipole interaction.

The CN groups assume varying angles with regard to the helical axis, guided by intramolecular repulsion and intermolecular attraction. In the unstrained state, two

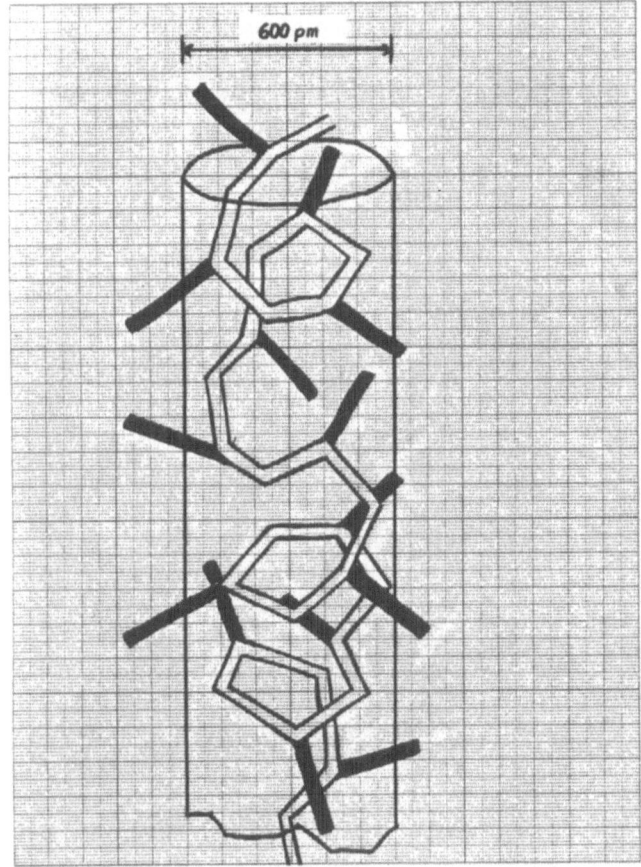

Fig. 1. Assumed rigid, irregularly helical conformation of the polyacrylonitrile molecule

adjacent macromolecules will tend to lower their energy by bringing together a maximum of groups that can interact favorably. The energy involved in the interaction of two dipoles, μ_1 and μ_2, at a distance r, is generally given by[15] [1]:

---

1 Equation (1) is exactly valid only if r is large compared to the charge separation within the dipoles[15]; for interacting CN groups from neighboring macromolecules this may be assumed to be mostly the case.

$$E = \frac{\mu_1 \mu_2}{r^3}(\cos \phi - 3 \cos \delta_1 \cos \delta_2) \tag{1}$$

(Angles and vectors as defined in Fig. 2.) The optimum value of interaction energy is gained if the dipoles are in the antiparallel position, where $\delta_1 = 90°$, $\delta_2 = 270°$, $\phi = 180°$, and hence:

$$E\!\downarrow\!\uparrow = -\mu^2/r^3 \tag{2}$$

In parallel end-to-end alignment, $\delta_1 = \delta_2 = 0$, $\phi = 0$, the energy gain would be $E_{\uparrow}^{\uparrow} = -2\mu^2/r^3$, i.e. at comparable r, it would be greater by a factor of two [cf. Eq.(2)].

For CN groups in polymer molecules, however, this orientation is sterically not feasible. Low molecular weight nitriles do not align in the end-to-end mode either, because the linear arrangement of the C–C≡N skeleton of these molecules would require a carbon atom in between two interacting CN dipoles; in the antiparallel position, the dipoles can come closer together. (Note that for parallel side-by-side position, $\delta_1 = \delta_2 = 90°$, $\phi = 0$, and $E\!\uparrow\!\uparrow = +\mu^2/r^3$, indicating maximum repulsion.)

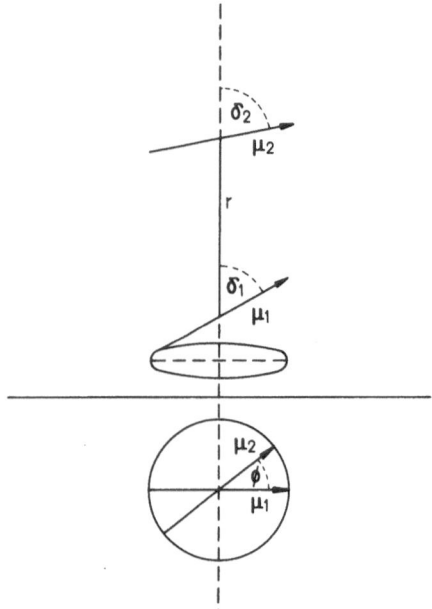

Fig. 2. Parameters involved in the interaction of two dipoles, μ_1 and μ_2

Evidently, only a few C≡N groups of the macromolecules can be assumed to be oriented in such a way that perfect antiparallel alignment with C≡N groups of next neighbors, at minimum r, becomes possible. However, energy gain will result also from interaction of CN groups oriented in a somewhat less ideal way.

2.3 Response to Stress

If a polymer is exposed to stress, a force develops within this polymer which provides resistance against deformation. S. L. Dart[16] has shown that, in the particular case of polyacrylonitrile fibers, this force is essentially due to the increase of internal energy (and not to entropy changes as in rubbers). S. Rosenbaum[14] has suggested that the increase of internal energy is caused by a short range straightening of the helical conformation of individual macromolecules, forcing them into a zig-zag conformation, against the repulsion of adjacent CN groups. This movement necessarily pulls apart also some of the intermolecular dipole pairs, or at least lifts them out of their position of lowest energy. After the shifting of polymer segments, new intermolecular interactions may become possible, but the overall situation will be less favorable than that without stress.

Below the glass transition temperature (T_g; ca 100 °C) the depicted mechanism is assumed to be the main response of polyacrylonitrile fibers to stress. It may account for an elongation of the fiber of up to 40%. As it may be expected, the extension is almost entirely reversible, if the stress is removed, and the kinetic conditions are provided permitting the macromolecules to restore the state of lowest energy (e. g. by steam treatment above T_g).

At $T > T_g$ additional mechanisms of response to stress become available. The macromolecules in a fiber are arranged in fibrils which, in turn, form a three-dimensional network. Several structural models have been used to visualize the deformation of the fiber during stretching. Knudsen and Fitzgerald[17a] assumed a network of rigid rods with elastic joints; Bell and Dumbleton[17b] replaced the rigid rods by springs in order to account also for the molecular deformations mentioned above (helical to zig-zag conformation). The effect of stretching is then attributed to deformation of the network as a whole, associated with changes in angle at the fibril juncture points, orientation of the fibrils in the direction of stretch, and eventually a slip of fibrils relative to one another. The very high break extension – up to 350% have been recorded[14] – appears to be a feature unique to fibers from polyacrylonitrile. Most probably, the intermolecular interactions of CN groups have their part also in this phenomenon.

In sufficiently stretched polyacrylonitrile (films or fibers), the individual untwisted macromolecules may be thought of as packed side-by-side, and laterally bonded by intermolecular dipolar interactions; these ordered zones are interrupted by amorphous regions where the intermolecular forces have not become effective. A periodic arrangement of more and less ordered regions has been found by a combined study of small and wide angle X-ray scattering, electron scattering and electron microscopic observations[18]. Within the more ordered regions the macromolecules build up a strongly disturbed three-dimensional crystal lattice. The elementary cell has been described by Hinrichsen and Orth[18a] as orthorhombic, with the dimensions a = 1.06 nm, b = 1.16 nm, c = 0.504 nm. Although the nonequatorial reflexes are very weak, they clearly indicate a periodicity in the c-direction (direction of the polymer chains). This periodicity is exactly twice in lenght as compared with the identity period of polyethylene, indicating a regular, untwisted zig-zag configuration of the carbon back bone in these zones. It was suggested that only syndiotactic chain sequences would be able to build up the crystalline regions. Acrylonitrile may, in fact, be expected to have a certain preponderance of sydiotactic growth steps during polymerization, due to the electrostatic repulsion of the polar side groups. The presence of longer sequences of syndiotactic placements, interrupted by single isotactic positions or short sequences of isotactic or atactic configurations has actually been deduced from a comparison of NMR, IR and Raman spectra of polyacrylonitrile[19]. The high perturbation in the crystalline regions would then be caused primarily by the natural distribution of lengths of the syndiotactic sequences, which forces sequences of other tacticity into the crystalline regions, disturbing the crystal lattice. However, the very fact that a defined melting point has been observed (cf. next section) confirms the partly crystalline character of stretched polyacrylonitrile.

2.4 Melting and Glass Transitions

Under ordinary conditions, polyacrylonitrile is not stable up to its melting point. Discoloration begins at, or even below, 150 °C, and in the temperature range of 180–380 °C, a number of exothermic reactions take place, including autocatalytic cyclization involving the CN groups, and decomposition leading to the production of gaseous products such as NH_3, HCN, nitriles, etc.[18a, 20]. Extremely high heating rates were necessary to determine the melting point of PAN by differential thermal analysis (DTA). At a heating rate of 40 °C/min[21], a small endothermic peak at 326 °C, superimposed on the large exothermic maximum of decomposition, was observed and ascribed to the melting of PAN. An unequivocal proof for the existance of a defined melting temperature was obtained by Hinrichsen[22] with the aid of even higher heating rates (over 1000 °C/min.). The samples (fibers) were immersed into a bath of molten metal, tempered there for a few seconds, and then cooled in ice-water mixture. Most of the exothermic decomposition reactions do not take place in such a short heating period. Samples tempered at temperatures below 320 °C presented clear X-ray reflexes indicative of partial crystallinity, and stereoscan measurements revealed unchanged fiber structure. Samples tempered at higher temperatures showed a molten surface in stereoscan, and the X-ray pictures indicated a thorough melting of the entire sample. Although this method does not permit the determination of the melting point more accurately than in the range (320 ± 5) °C, it provides a clearcut evidence for the fact of melting.

The melting temperature for PAN is extremely high as compared, e. g., with that of isotactic polystyrene (230 °C) or high density polyethylene (137–140 °C). Generally, a high melting point, T_m, can be caused by a high heat of fusion, ΔH_f, and/or by a low entropy of fusion, ΔS_f:

$$T_m = \Delta H_f / \Delta S_f \tag{3}$$

Kriegbaum and Takita[12] reported a heat of fusion of 5.0 kJ/mol (per repeat unit), a value which is below that of polyethylene (6.7 kJ/mol). Since it appeared difficult of reconcile strong attractive forces with the chemical structure of polyethylene, these authors pointed out that it appeared unlikely to them that strong intermolecular forces would contribute to the high melting point of PAN. And yet relatively strong, intermolecular dipole-pair bonding had to be taken into account in order to interpret adequately the response of PAN to stress (cf. preceding section). The solution to this apparent enigma lies probably in an undervaluation of the attractive dispersion forces in polyethylene. In the tightly packed arrangement of crystalline polyethylene, with low intermolecular distances, the dispersion forces may well be of the same order of magnitude as dipole interactions in more polar molecules[15]. In PAN, on the other hand, the protruding nitrile groups cause relatively large intermolecular distances. The dispersion forces, which decline with the sixth power of the distance, may be assumed to be very small under these conditions. But dipole-dipole attraction between the nitrile groups of neighboring macromolecules, as described in the preceding section, brings the heat of fusion in the vicinity of that of polyethylene.

At any rate, however, the intermolecular interactions are certainly not strong enough to account alone for the very high melting point. The entropy of fusion, as determined from Eq. (3), gives valuable insight into the changes of ordering occurring during melting. Expressed as entropy of fusion per single chain bond, a value of 4 J/°C has been given for PAN[12], well below the usual range of 6–8 J/°C for polymers like polyethylene, polychloroprene, poly(hexamethylene terephthalate), etc. This fact indicates that few degrees of freedom are gained if PAN passes from the crystalline to the molten state, and thus corroborates the view of a stiff, stretched molecule, forced into a helical conformation by steric and electrostatic repulsion of the CN group, even in the molten state. Hence, molecular stiffness may safely be assumed to be the main reason for the high melting point of polyacrylonitrile.

No clear picture can be gained from the literature concerning number and exact temperature range of glass transitions of PAN, between room temperature and the melting point. Transitions almost everywhere between 39 °C[23] and 180 °C[24] have been reported. Several methods have been applied to tackle the problem: birefringence[25], infrared dichroism[26], wide angle X-ray[27], as well as dynamic mechanical[28] measurements. Evidently the observed transitions depend not only on the method applied, but also on sample form (film or fiber), morphological state, thermal treatment, relative humidity, solvent content, and other factors. Most authors appear to agree on two major transitions, one at about 80–100 °C, and the other in the vicinity of 140 °C. Commonly it is assumed that the thermal movement overcomes the weak dispersion forces in the lower transition region (onset of backbone mobility), whereas the intermolecular dipole-dipole interactions remain intact at this point. This view is corroborated by the fact that the Young modulus of PAN fibers decreases only by a factor of $\simeq 35$ in this region, and not by the usual factor of $\simeq 1000$ observed with other polymers at their glass transition[29]. As the temperature increases further, a gradual decrease in modulus occurs. This behavior indicates that in the lower of the two major transition regions PAN does not transform from the glassy to a fully rubbery state, but to a state which is somewhat comparable to a highly crosslinked rubber[30]. The higher transition region is then ascribed to a loosening of the intermolecular dipole-dipole interactions.

Interestingly, the lattice spacings of the more ordered regions of stretched PAN increase with temperature in a discontinuous way, in the range of 80–120 °C, as gathered from wide angle X-ray scattering[27]. This was taken as an evidence that the motions related to the transitions in this temperature range are localized in the more ordered regions. Since the molecular structure in these regions is more disturbed than in other semicrystalline polymers, it appears plausible to attribute a certain mobility to chain segments located in the more ordered zones, even below the melting point. If this hypothesis is correct, corresponding motions in the less ordered regions may be expected to freeze at lower temperatures, giving rise to additional transitions.

The addition of small amounts of comonomer, such as methylacrylate or vinylacetate (Sect. 2.5) generally causes a merging of the two major transitions, giving rise to one broad transition occuring at 75–100 °C.

2.5 The Consequences of Copolymerization

A "good" textile fiber is a complex system of crystalline and amorphous zones, which is expected to combine the beneficial properties of each. Generally, high tensile strength, rigidity and stability of shape are anchored in the crystalline regions, whereas the amorphous regions provide the fiber with elasticity, and with segment mobility necessary for efficient dye diffusion. Stretched fibers from PAN homopolymer have excellent tensile properties, but the properties depending on the amorphous phase are not satisfactory. The reason is a relatively high degree of ordering and intermolecular CN interaction even in the "less ordered" zones. Consequently, most commercial fibers from acrylonitrile are actually made from copolymers, containing 3–7% of a vinyl monomer, mostly methyl acrylate, methyl methacrylate or vinyl acetate. The introduction of such small amounts of a comonomer greatly enhances the internal mobility of polymer segments, reducing the sequences of acrylonitrile molecules capable of interacting with neighboring sequences. The dye diffusion is substantially improved, and the low temperature brittleness is decreased.

2.6 The Effect of Polar Additives

Low molecular weight substances offering a high dipole moment, such as nitriles, amides, nitro compounds, etc., (Table 3) have a strong plasticization effect on poly-

Table 3. Dipole moments of some selected compounds[2]

Compound	Dipole moment (Debye)
Acetonitrile	3.72
Propionitrile	4.02
Benzonitrile	4.18
Dimethylformamide	3.82
Dimethylacetamide	3.81
Dimethylsulfoxide	3.96
Benzamide	3.6
Aldehydes and ketones	2.7–3.7
Organic nitro compounds	3–5
Urea	4.6
Nitric acid	2.16

acrylonitrile fibers, i. e., in their presence the resistance of the fiber to an imposed stress is decreased[31]. (Hydrogen-bonding additives, which have a similar effect, are treated in Sect. 3.6).

What can be expected, if such molecules are introduced into the fiber? The more mobile additive molecules have a greater chance to approach PAN $-C \equiv N$ dipoles in the most favorable direction [cf. Eq. (2)]. By such action, the intramolecular repulsion of adjacent $-C \equiv N$ groups within the polymer molecules will be partly "neutralized", with the result of less resistance to an imposed stress. At the same time, part of the intermolecular dipole-dipole interactions between polymer molecules will be replaced by interactions of polymer CN groups with dipoles of the additive, which again results in reduced tensile strength. On the other hand, the presence of the "plasticizing" compound evidently provides increased mobility of individual segments of the polymer molecules. As a consequence, the glass transition temperature is lowered, and the dyeability is improved.

In the extreme case, where all polymer CN dipoles interact with the dipoles of a liquid, strongly polar "additive", dissolution of the polymer is likely (e. g., with dimethylformamide, dimethylacetamide, or dimethylsulfoxide).

3 Hydrogen Bonding

3.1 Some General Aspects of Hydrogen Bonding

Extended reviews concerning all aspects of hydrogen bonding are available in the literature[32, 33]. Only a few relevant features will be recalled here.

A hydrogen bond of the type we have to consider here (i. e. a non-ionic hydrogen bond) generally involves a neutral, electronegative atom X, such as N or O, offering a lone pair of electrons, and a neutral, though moderately polar, short and strong Y–H bond, such as N–H or O–H, as the electron acceptor:

$$X \cdots \cdots H{-}Y$$

Experimental[15] and theoretical[32, 33] studies have shown that the energy gained by hydrogen bond formation is in the range of 5–50 kJ/mol, depending on X and Y. Data obtained by Kollman and Allen[33], using ab initio self consistent field molecular orbital (SCF MO) theory, are given in Table 4. It may be seen that, for the same electron acceptor, the hydrogen bond strength decreases with varying electron donor in the series $N > O > F$. This series parallels the availability of the lone pair orbitals of the donor atoms, as borne out by the ionization potentials of these orbitals, viz., 10.16 eV for NH_3, 12.61 eV for H_2O[34], and 16.38 eV for HF[35].

For the same electron donor, on the other hand, the hydrogen bond strength decreases, as the electron acceptor changes, in the series $F{-}H > O{-}H > N{-}H$, which parallels the positive charge on the proton, as borne out by the electronegativity of fluorine (4.10), oxygen (3.50), and nitrogen (3.07)[36]. The trends outlined in Table 4 follow also from calculations of Umeyama and Morokuma[32], although the

Table 4. Calculated H-bond energies and heavy atom distances[33]

Electron Donor	Electron Acceptor		
	NH_3	H_2O	HF
H-bond energies (kJ mol^{-1})			
NH_3	11.3	24.5	48.9
H_2O	9.6	22.1	39.3
HF	5.4	12.5	19.2
Heavy atom distances (pm)^a			
NH_3	349 (342)	312 (303)	275 (304)
H_2O	341 (303)	300 (264)	277 (265)
HF	345 (304)	308 (265)	288 (266)

^{a} In parentheses: sum of heavy atom ionic radii (N^{3-}: 171, O^{2-}: 132, F^{1-}: 133 pm)[2]).

individual binding energies are somewhat higher, due to a different SCF–MO approximation.

Table 4 shows also the (calculated) minimum energy distances between the involved heavy atoms. Despite the presence of the hydrogen in-between the two heavy atoms, the distances are only slightly larger (or even smaller) than the sum of the heavy atom ionic radii. The proton is pulled towards the electron donor X (as indicated by a decrease of the H–Y stretching frequency[14]), and plunges somewhat into the electron cloud of the latter; this approach is the greater, the better available are the lone pair orbitals of X, and the more positively charged is the proton. The heavy atom distance does not change essentially for the same electron acceptor, as the donor atom is varied. However, the difference between the heavy atom distance and the sum of the ionic radii (which increases in the series $N < O < F$) clearly reflects the better availability of the lone pairs in the reverse series (greater attraction). The considerable reduction of heavy atom distance on going from left to right in Table 4 reflects both, the decrease in ionic radii, and the increasing attraction due to the enhanced electropositive character of the proton.

Hydrogen bonds tend to have a linear X . . . . H–Y arrangement. In former times, hydrogen bonding has been thought of as entirely electrostatic in nature, and electrostatic models were indeed able to reproduce observed H-bond energies and geometries. However, the electrostatic models could not account for certain other experimental findings. Thus, the intensity of the H–Y stretching band increases, and that of the in-plane bending mode decreases upon H-bond formation; moreover, there is no correlation between H-bond strength and the dipole moment of the electron donor[37]. Application of quantum mechanics, in particular of the Morokuma component analysis[32], to the problem of hydrogen bonding, shed new light onto the energies involved. This analysis permits one to break down the total SCF interaction energy into electrostatic, polarization, charge transfer, and exchange repulsion components. Polarization (i. e., charge redistribution within the constituent fragments of an H-bond) generally plays only a small role. Charge transfer between the fragments and exchange repulsion, on the other hand, are often of the same order of magnitude as the electrostatic contribution, but their influences tend to cancel[32, 38a]. This is the reason why the electrostatic models were successful, and are still very convenient for understanding and predicting many aspects of hydrogen bonding[38b]. The near cancellation of the two quantum mechanical effects mentioned above may be rationalized, in a rough approximation, in the following manner: charge transfer from the lone pair orbital of the electron donor X to the empty antibonding σ^*-orbital of H—Y leads to an energy gain, whereas the interaction of the lone pair with the filled bonding σ-orbital of H—Y causes repulsion. Increased availability of the lone pair enhances both effects. The charge redistribution by charge transfer and polarization explains the spectroscopic findings.

Intermolecular hydrogen bonding between CN nitrogens and tertiary hydrogens has been considered previously as the main interaction responsible for the fiber-forming capacity of polyacrylonitrile (for references see[3]). Although such hydrogen bonds may be present to a certain extent, their importance is now considered small compared with the intermolecular dipole-dipole interaction of nitrile groups, for the following reasons. Whereas the energy gained by dipole-dipole interaction is in the range of 20–30 kJ/mol (Table 1), the energy of a hydrogen bond of the type N . . . ., H—C may be assumed to be in the range of 4–5 kJ/mol[32]. Moreover, Saum[3] has shown that the remarkable increase of the boiling point of nitriles, as compared with hydrocarbons of the same size, is nearly the same for normal, secondary, or tertiary butyronitrile; since the increase of boiling point is evidently due to association, and since tertiary butyronitrile does not have a hydrogen at the α-carbon atom, hydrogen bonding can be excluded as the cause of the association.

Water is an almost omnipresent hydrogen bonding partner for acrylic fibers. In order to understand its interaction with the polymer, it is useful to study its structure and energetics on a molecular basis.

3.2 The Structure of Water

The structure of water is determined largely by the ability of the H_2O molecule to form hydrogen bonds. In principle, every molecule may take part in four such bonds: with the two lone pairs of electrons located at the oxygen as electron donor sites, and with the two protons as electron acceptor sites. In contrast with other small molecules such as dinitrogen, carbon dioxide, methane, ammonia, hydrogen cyanide, etc., water has in fact anomalously high melting, boiling, and critical temperatures, which characterize it as a strongly bound many-body system. Luck[39] has estimated, using the IR overtone method, that between room temperature and 100 °C only some 10–20% of the OH groups of water are not involved in H-bonding. This finding makes it improbable that a noticeable amount of monomeric water molecules, with both OH bonds free, are present in liquid water.

Computer simulations (Monte Carlo method[40, 41] and molecular dynamics calculations[42, 43]) have been used to analyze the structure of water. According to these studies, water is made up of a continuous distribution of associates. The pair interaction between nearest neighbor molecules should not be thought of as either "made" or "broken" hydrogen bond, but rather as a continuum of more or less bent and stretched H-bonds with a continuous spectrum of energy environments. Based on computer simulation, the ratio of particles with association number $K < 4$ to larger particles is 70/30, with a maximum at $K = 4$[41]. Experiment based estimates (IR overtone method[39]), on the other hand, come up with a cluster size of 240 water molecules at room temperature, and of about 40 molecules at 100 °C.

Nevertheless, minimum energy geometry and binding energy has primarily been calculated for dimers of water. More recent literature data are summarized in Table 5. With some scattering (due to the use of varying basic sets in the calculations) the binding energy is found in the region of 20—25 kJ/mol, slightly higher than the binding energy in dipole-dipole interaction of nitriles (Table 1). The close agreement with experimental data (Table 5, lower part) should be considered as accidental, if the present view of the structure of water is correct. Presumably, the effect of higher aggregation is compensated for by the presence of very weak bent and/or stretched hydrogen bonds.

The lowest energy geometry of the dimer $(H_2O)_2$ is represented in Fig. 3. Most authors agree in that the hydrogen bond is essentially linear (bond angle $O^1H^2O^2 = 180° \pm 10°$). The xz plane is a symmetry plane for the dimer, i.e. the bisector T of the bond angle $H^3O^2H^4$ lies in this plane, with an orientation angle θ to minimize repulsion between hydrogens. Some discrepancy exists concerning θ, for which an experimental value of $(60 \pm 10°)$ has been reported[53], whereas most calculations resulted in somewhat lower angles. However, Kollman[39a] has pointed out that more extented basis sets (as compared with those used till now in the calculations) would tend to give higher values of θ. Table 5 also indicates that the optimal distance between the two oxygen atoms, $R_{0-0'}$ is very close to 300 pm.

Table 5. Calculated parameters for minimum-energy dimers of water, and some experimental data

Calculation method[a] or experimental basis	Binding energy (kJ mol^{-1})	R_{0-0} (pm)	θ (deg)	Ref.	Year
HF	20.9	300	40	44)	1970
HF	22.1	300	–	33)	1971
HF	26.6	274	57	45)	1972
HF	19.0	301	–	46)	1973
HF	33.8	285	52	47)	1974
CI	25.3	290	–	48)	1975
HF	30.9	285	37	49)	1975
CI	23.5	292	–	50)	1976
HF	32.6	288	37	32)	1977
Heat of vaporization	21.7	–	–	51)	1931
Heat of sublimation	24.7	–	–	52)	1933
Microwave spectrosc.	–	298	60 + 10	53)	1974

[a]   HF: ab initio (all electron) Hartree Fock calculation
      CI: calculation including configuration interaction.

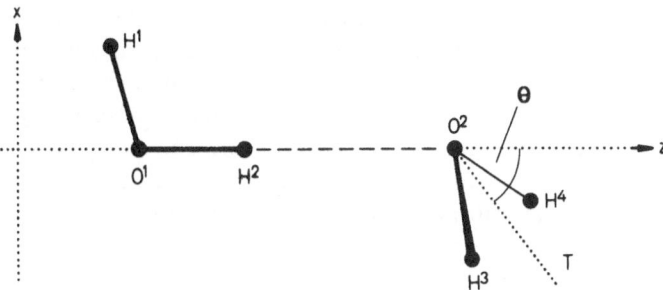

Fig. 3. Lowest-energy dimer structure of water

Stillinger[43] has studied possible motions out of the equilibrium position of the water dimer, with regard to energy requirements. Motions that tend to change R_{0-0} from its optimal value, or to move the electron acceptor (H^2 in Fig. 3) off the oxygen-oxygen axis, are costly in terms of stabilization energy. But rotation of the acceptor molecule $H^1 O^1 H^2$ about the linear $O^1 H^2 O^2$ axis, as well as change of the orientation angle θ, are relatively easy (Sect. 3.4).

3.3 Nitrile-Water Interaction

Organic nitriles associate to dimers by dipole-dipole interaction of their nitrile groups (Sect. 2.1); water molecules, on the other hand, associate by formation of hydrogen bonds (preceding section). However, if water and nitrile (in particular acetonitrile[54]) are mixed, heterodimers and -trimers are formed:

R–C≡N . . . H–O–R   R–C≡N . . . H–O–H . . . N≡C–R

These complexes have been investigated by observing variations of the infrared stretching frequency of the OH group of the water molecule. If the molar ratio of water to nitrile is < 0.1, the heterotrimer is observed; if a small amount of nitrile is added to water, the heterodimer is predominant.

In the case of polyacrylonitrile, the formation of "heterotrimers" could evidently lead to crosslinking. Actually, the gel-forming effect of small amounts of water on concentrated solutions of PAN is known[55], and may be caused by this phenomenon.

There has been some discussion as to whether the water-nitrile interaction involves the lone pair orbital of the nitrogen, or the π-orbitals of the C≡N bond[56, 57]. The experimental ionization potential of the former is 13.14 eV, that of the latter is 12.21 eV[58], hence the bonding π-orbitals are, evidently, the highest occupied molecular orbitals of the nitrile. Nevertheless, calculations of Del Bene[56] have shown that heterodimers with the π-orbitals acting as the electron donor are not equilibrium structures on the intermolecular potential surface, i. e., even if formed initially, such dimers would eventually be converted into heterodimers in which the hydrogen bonding involves the lone pair of electrons at the nitrogen.

Experimental evidence confirms this view: the interaction of water with the nitrile group (of acetonitrile) leads to an increase of the stretching frequency of the C≡N bond from 2254 to 2261 cm^{-1} ($\Delta \nu_{CN}$ = 7 cm^{-1} [59]), indicating a moderate stabilization of this bond. The hypsochromic shift of the C≡N stretching frequency is diagnostic for interaction of the nitrogen lone pair electrons with an electron acceptor, and has been observed in numerous other complexes; interaction of the C≡N bonding π-electrons with an acceptor, on the other hand, leads to a bathochromic shift of ν_{CN} (Sect. 4.1).

Comparing the ionization potentials of the lone pair orbitals of nitrogen in acetonitrile (13.14 eV) and of oxygen in water (12.61 eV[34]), it may be anticipated that the bonding energy of the heterodimers and -trimers should be only slightly lower than that of H$_2$O . . . HOH. Thus, water dimers, nitrile dimers and water-nitrile heterodimers appear to have all similar stabilization energies. That means that the combined process of dissociating a water dimer and a nitrile dimer, and forming heterodimers thereof, certainly does not consume a great deal of energy:

$$(H_2O)_2 + (RCN)_2 \rightleftarrows 2\ RCN . . . HOH \tag{4}$$

Possibly, there is even a small energy gain to be expected, if the stabilization energy of the heterodimer is slightly larger than that of the dipole-dipole interaction in the nitrile dimer.

3.4 The Effect of Water in Acrylonitrile Polymerization

Acrylonitrile (AN) is a particular case among vinyl monomers regarding the solvent dependence of the propagation rate constant (k_p) in homogeneous free radical polymerization. Whereas in dimethylformamide and dimethylsulfoxide k_p has values which might be expected from a comparison with other monomers and from the reactivity of the growing radical (400 and 1910 l mol^{-1}s^{-1}, respectively, at 25 °C[60]), the propagation rate constant is considerably higher in water (2.8 x 10^4 l mol^{-1}s^{-1} [60]), and almost approaches values reported for ionic polymerizations[61]. Summarizing work concerning this phenomenon up to 1963, Dainton[62] indicated that the rate constants in water and in organic solvents could not be reconciled.

It appears plausible now to assume the water-nitrile interaction discussed in the preceding section responsible, in one way or the other, for the "water effect" in AN polymerization.

However, alcohols also form hydrogen bonds of the same type with nitriles; thus, a hypsochromic shift of 8 cm^{-1} of ν_{CN} has been observed with phenol[63]. And yet the polymerization rate of acrylonitrile in alcohols as solvent is comparable to that in dimethylformamide[64]. One has to conclude that the mere fact of this type of interaction alone can certainly not account for the "water effect" in AN polymerization.

We recently suggested[65] that the important peculiarity of the water molecule resides in its ability to form simultaneously two hydrogen bonds, e. g., one with the ultimate C≡N group of a growing polyacrylonitrile radical, and one with a monomer molecule. This arrangement not only would increase the effective local concentration of monomer in the neighborhood of the radical (cf. the "template effect" in coordination catalysis[66]), but also might activate the radical as well as the monomer by decreased delocalization of the free electron of the radical[67] and of the electrons of the vinyl double bond, on stabilizing the $C \equiv N$ bond.

We have to consider the structure of water in order to estimate whether the vinyl double bond of an acrylonitrile molecule, hydrogen-bonded to a water molecule via its nitrogen lone pair orbital, may come into reaction distance (say 300 pm = 3 Å) to the radical site of a growing chain coordinated to the same species. For an estimate of the geometry of the possible reaction complexes we use literature data of bond lengths and bond angles, as summarized in Table 6.

Figure 4 shows polymer radical and monomer coordinated to a single water molecule. A simple, straightforward trigonometric evolution, using the data of Table 6, indicates that both hydrogen bonds would have to be bent about 40 ° out of their lowest energy (linear) position, in order to bring the reacting ends into the required distance. This movement is relatively costly in terms of stabilization energy (Sect. 3.2), and might therefore be prohibitive.

In the water dimer (Fig. 3), the situation is somewhat more favorable. As mentioned in Sect. 3.2, the orientation angle θ is ca 60 °[2]; variation of this angle, as

Table 6. Bond lengths and bond angles in polyacrylonitrile, acrylonitrile, water, and hydrogen bonds

Bond	Bond length (pm)	Bond angle	Degrees	Ref.
		$P_n-C^1H_2-HC^{2*}$ $-C^3N^1 + C^4H_2 = C^5H-C^6N^2 \longrightarrow P^*_{n+2}$		
C^2-C^3	143	$C^4-C^5-C^6$	122.4 °	68)
C^5-C^6	143	$C^2-C^3-N^1$	180 °	
C^4-C^5	134	$C^5-C^6-N^2$	180 °	
C–N	116			
		H–O–H		
H–O	95.7	H–O–H	104.5 °	43)
		O . . . H–O		
O . . . H	204	O . . . H–O	180 °	43)
		N . . . H–O		
N . . . H	212	N . . . H–O	180 °	33)

2 In a preceding short communication[65], we used a value of $\theta \simeq 35$ °, as calculated by Umeyama and Morokuma[32]. In the meantime, the experimental value of $\theta \simeq 60$ °[53] came to our attention. With regard to a brief discussion of calculated values of θ see Section 3.2.

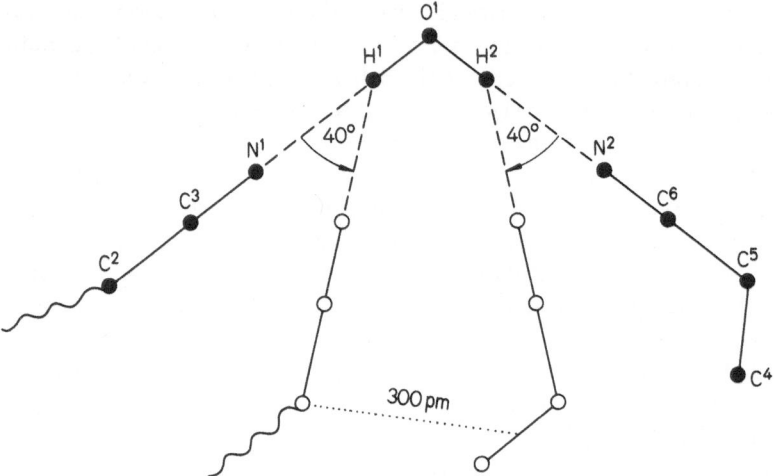

Fig. 4. Growing polyacrylonitrile radical $P_nC^1H_2-\overset{*}{C}{}^2H$ and monomer $C^4H_2=C^5H$, both coordi-
nated to a single water molecule $\underset{C^3N^1}{|}$ $\underset{C^6N^2}{|}$

well as a rotation of either molecule about the $O^1H^2O^2$ axis is not very costly. This
brings us to Fig. 5 where molecule $H^1O^1H^2$ is rotated by 180 ° about the hydrogen
bond axis, as compared with Fig. 3. As shown in Fig. 5, an increase of the orientation angle θ by $\simeq 40$ ° would be required to bring the reactants into reaction
distance.

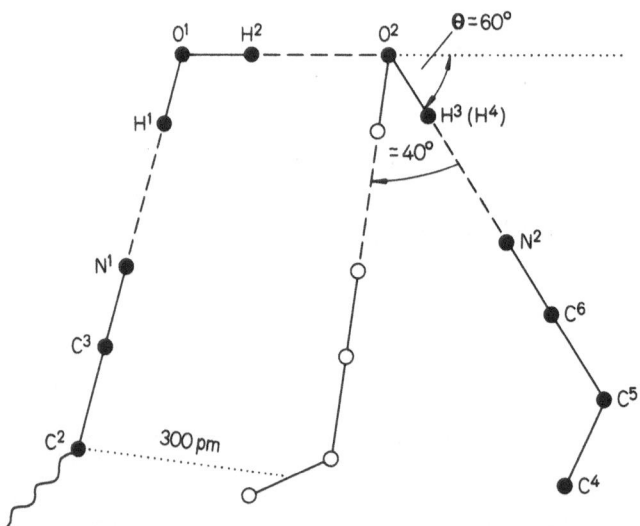

Fig. 5. Growing radical and monomer coordinated to a water dimer

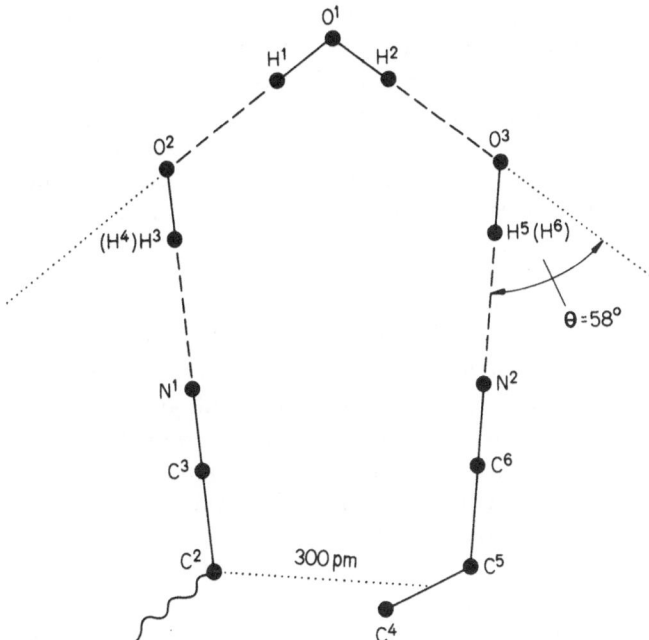

Fig. 6. Growing radical and monomer coordinated to a water trimer

The most favorable situation is given with the water trimer (Fig. 6). An orientation angle of 58 ° is necessary to bring the radical (C^2) and the double bond (C^4–C^5) into reaction distance. This angle is in perfect agreement with that found for the most stable form of interaction ($60° \pm 10°$)[53].

The water trimer sketched in Fig. 6 is one of three possible trimer structures[43]. The central $H^1O^1H^2$ molecule acts as double electron acceptor. Alternatively, the central water molecule could act as double electron donor, with the two lone pairs of electrons at O^1, or it could use one lone pair and one proton to build up a "sequential" trimer. However, due to the nearly tetrahedral arrangement of protons and lone pairs on the oxygen atom in each water molecule, and to the orientation angle of $\simeq 60$ °, each of the trimer structures would permit a favorable approach of polymer radical and monomer molecule, as depicted in Fig. 6 for one of them.

A similar arrangement of three vicinal water molecules can be assumed to be present also in the higher aggregates of water. Thus it can be concluded that water may well act as a "template", bringing together — and at the same time activating — the growing polyacrylonitrile radical and the monomer molecule. This behavior may easily explain the extremely high chain propagation rate constant in water. It has been pointed out[65] that the capability of water to act as a template might also be one important aspect of the crucial role of water in biochemistry.

3.5 The Effect of Water on Polyacrylonitrile Fibers

The strong interaction between H_2O and the nitrile group has a profound influence on a number of properties of polyacrylonitrile and fibers thereof.

It has been reported[69] that the addition of small amounts of water to acrylonitrile polymer ($H_2O/CN < 1$) lowers the melting point sufficiently to permit melt spinning at $\simeq 180\ °C$ and 40–70 atm. A single phase is formed when the finely powdered mixture is heated under the indicated pressure. The stretching frequency of the polymer CN groups at 2234 cm^{-1} disappears; instead a new band at 2050 cm^{-1} is observed. However, this large bathochromic shift is atypical for simple nitrile-water interaction by hydrogen bonding (Sect. 3.3). Presumably some kind of reversible hydration of the CN group takes place under the applied conditions of high temperature and pressure.

Under normal pressure, polyacrylonitrile is strongly plasticized by water, i. e., its resistance to an imposed stress is greatly decreased, in particular in hot water. However, unlike dimethylformamide or dimethylsulfoxide, water is not able to dissolve polyacrylonitrile, although the energy of interaction with the nitrile groups should not be very different for water as compared with the two solvents. Water has been characterized as a strongly bonding many-body system (Sect. 3.2). It appears reasonable to assume that the high degree of aggregation of the water molecules seriously reduces their diffusion within the polymer.

Nevertheless, the migration of water into the fiber is sufficient to produce the above-mentioned plasticization effect. The chain mobility is increased, as indicated by a decrease of the glass transition temperature by 35–50 °C[16, 17]. This is a very fortunate fact, because dyeing of the fibers is possible only above the Tg, where the increased polymer segment mobility permits dye diffusion within the fiber[3]. For the commercial polyacrylonitrile copolymer fibers, the Tg in water lies in the region of $\simeq 80\ °C$[70], so that dyeing at, or slightly below, the boiling point of water becomes feasible.

However, acrylic fibers have to go through a relaxation process, before satisfactory dyeing rates can be achieved — and here again water plays an important role. The fiber forming process (comprising spinning, stretching and drying) leaves the fiber with too high a degree of ordering to permit sufficient penetration of dyestuff. A thermal treatment, mostly with superheated steam, brings about a more porous structure. Sotton et al.[72] have shown that the internal fiber morphology, and particularly the pore distribution, is quite different if the thermal treatment is made under nitrogen, or with water vapor. Figure 7 shows schematically these differences. Fibers relaxed in dry heat tend to have a concentric array of porous zones, the core of the fiber remaining often clear of voids; water vapor treated fibers, on the other hand, appear to have crevices and splits which lead from the periphery right into the inner zones of the fiber. Actually, the latter fibers have considerably higher initial dyeing rates than the former[73]. (This difference is, naturally, levelled down after longer residence times in the hot dyeing bath.)

Wearing comfort is another area where nitrile-water interactions are of utmost importance. Wearing comfort requires — among other things — easy wetting of, and good moisture transport within, the fabric or knit used for the manufacture of a garment. Body moisture should be absorbed rapidly and steadily, and transported

---

3 Dye diffusion within the fiber is assumed to occur by migration of the dye molecule or ion in the plasticized polymer, rather than by diffusion of bulk water solution of the dyestuff[71].

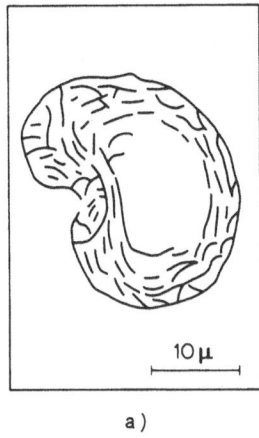

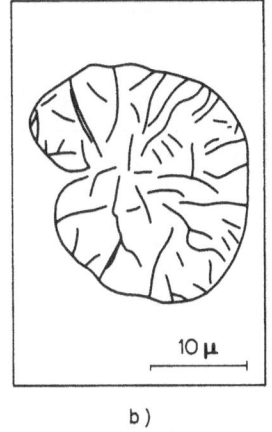

Fig. 7. Schematic representation of the cross-section of heat-treated (130 °C) acrylic fibers (from AN/methylmethacrylate copolymer); **a)** under nitrogen, **b)** with water vapor. (Adapted from electron-micrographs of Sotton et al.[72])

a)   b)

through the fabric, so that the moisture evaporates quickly. Although the structure of fiber, yarn and fabric has doubtlessly an influence on these important properties, they are also intimately connected with the net attraction between the fiber surface and water. Acrylic fibers are, in this respect superior to wool, nylon, and polyester, and similar to cotton and viscose, as estimated by the contact angle formed between the solid surface and water[4][74, 75], see Table 7. Contact angles have been criticized as not bearing a simple relationship to the energetics of interaction between a solid and a liquid[76]. However, as long as two solids are compared using the same wetting liquid (water in this case), a difference in contact angle does tell which solid has the greater attraction for the liquid (the smaller the contact angle, the greater the attraction, cf. Fig. 8).

Experimentally, the measurement of contact angles is somewhat tricky, hence absolute values should not be considered too seriously. In particular, there is generally a difference between the angle found when the liquid advances over the solid, and the angle found when the liquid recedes[75]; this is because the previous condition of the surface affects its wettability. But a series of measurements made under the same conditions certainly indicates the right ordering of fiber materials with regard to interaction with water.

An interesting practical test concerning comfort of acrylic fibers has been reported recently[77]: A group of 69 basketball players were equipped with one or more socks made of acrylic fibers on one foot, and the same number of socks made of cotton or wool on the other. The class of fiber was not identified to the players. The result was that 59 out of the 69 athletes preferred the socks from acrylic fibers versus those from cotton or wool, because they kept their feet drier, and felt softer.

In our context it is important that it was clearly shown in the report[77] that the favorable response to acrylic fibers was due to a water transport phenomenon, and not to water retention. (Wool and cotton are quite superior in this respect, however, above a certain minimum limit this property appears to be adverse to comfort, due to the sensation of clamminess.)

---

4   This statement can, of course, be invalidated by surface treatments.

Table 7. Contact angles between water (advancing) and fiber, for several polymeric materials[74]

Fiber material[a]	Contact angle (degree)
Polyethylene, polypropylene	92
Wool	91
Polyester	75
Nylon	70
Polyacrylonitrile	48
Cotton	47
Viscose	40
Treated wool[b]	26–78

[a] Extracted with alcohol and ether, conditioned at 22 °C, 67 % relative humidity.

[b] Varying treatments; e.g. 16 % hydrochloric acid, sulfuryl dichloride or alcoholic potash.

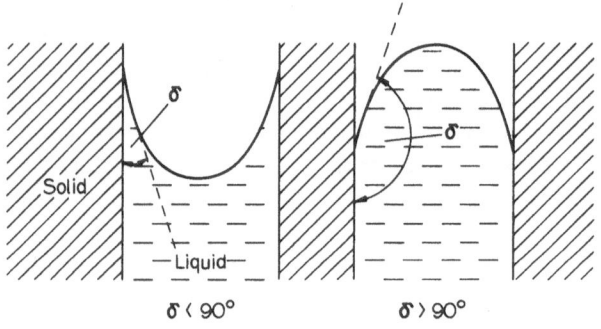

$\delta < 90°$ $\delta > 90°$

Fig. 8. Definition of contact angle in a capillary

From the above it appears clear that the moisture transport takes place at the surface of the fiber (solid-liquid interface). The capacity of water retention, on the other hand, appears to reside primarily in macrovoids within the fiber. Actually, acrylic fibers with high water retention can be obtained, according to a recent patent[78], by spinning a fiber from a blend of an acrylic copolymer and a highly soluble component (e. g., glycerine), which later is washed out in the further treatment, leaving behind voids and capillaries which provide the high water retention.

3.6 Hydrogen Bonding Additives and Solvents

The plasticizing action of several hydrogen bonding, low molecular weight compounds such as methanol, ethanol, benzyl alcohol, ethylene glycol, phenol, aniline, and acetic acid has been well established. In one series of experiments, the load-extension curve has been determined for polyacrylonitrile fibers immersed into the respective liquids[31]. In another, hydroxy or amino compounds have been added to water, and the variation of the glass transition temperature was measured[79]. The Tg in water-benzyl alcohol mixtures was found to be lower than in either water or the alcohol alone, indicating a certain additivity of the effects if more than one plasticizing agent are present. The addition of benzyl alcohol to a dye solution in water increased the rate of dye diffusion within the fiber[79], again indicating additional chain mobility.

Note that plasticization is not restricted to hydrogen bonding liquids; the effect of compounds with high dipole moments such as nitriles, amides, nitro compounds, which act via dipole-dipole interaction with the nitrile groups of polyacrylonitrile, has been treated in Sect. 2.6.

The dissolution of polyacrylonitrile in conc. H_2SO_4[60] is, most probably, primarily an effect of strong hydrogen bonding; it is, however, accompanied by some saponification of the nitrile groups. Concentrated nitric acid also dissolves PAN, and has even been used as a solvent for fiber spinning[80]. The dissolving power of conc. HNO_3 probably resides in the combined action of hydrogen bonding and the relatively high dipole moment (cf. Table 3).

4 Electron-Donor-Acceptor Complexes

4.1 Complexes with Model Substances

Organic nitriles, in particular acetonitrile, form stable electron-donor-acceptor (EDA) complexes with a number of Lewis acids such as $BX_3(X = Cl, Br, I)$, $SnCl_4$, $AlCl_3$, and with many transition metal halides, the nitrile group acting as the electron donor[81].

The donor character of the nitrile group can arise either by donation of electrons from the π-orbitals of the CN triple bond, or from the lone pair localized at the nitrogen atom. From photoelectron spectroscopy (acetonitrile) it has been concluded that the degenerate pair of bonding π-orbitals is the highest occupied level (-12.21 eV), and that the lone pair orbital is more stable (-13.14 eV)[58]. SCF–MO calculations have confirmed these assignments[82]. Hence it might have been expected that an "acetylene-like" coordination via the π-orbitals would be the predominant mode. However, with the exception of some particular chelating dinitriles, which will be discussed below, it appears that in general the lone pair orbital at the nitrogen atom is involved in coordinative bonding of nitriles. Evidently the polarity of the CN bond is primarily responsible for this fact: a greater energy gain may be anticipated if the electron acceptor approaches the CN group at the negative end of the dipole. The strongly directional lone pair orbital (sp), which presumably protrudes further into space than the π-orbitals, may then serve as a guide to bond formation.

A few X-ray crystallographic studies of complexes containing CH_3CN, or dinitriles of the type $NC(CH_2)_nCN$, have been reported. The $-C\equiv N \ldots A$ arrangement, with A = electron acceptor, is generally linear as shown below for A = BF_3, as an example[83]:

$$H_3C \text{---} C \equiv N \ldots \ldots BF_3$$
$$\text{---}144113164$$
$$\text{---}pmpmpm$$

Amazingly, the stretching frequency of the CN bond, ν_{CN}, is shifted to higher values in the complexes, compared to that of the free nitriles (see Table 8). It has been argued that this hypsochromic shift is mainly due to a kinematic coupling of the $C \equiv N$ and N . . . A bonds during exitation of the CN normal mode[90]. However, Purcell and Drago[85] have shown that the latter effect can account for only a part of the frequency shift, and that the force constant of the CN bond is actually increased; both the bonding π- and the bonding σ-orbitals are more stable in the complexes than in the isolated nitrile molecules. In accordance, the CN bond length was found to be slightly shorter in the complexes (cf. Table 8).

Based on data like those shown in Table 8, it has become customary to consider an increase of the stretching frequency, ν_{CN}, as diagnostic of complex formation via the lone pair orbital at the nitrogen atom, even if no single crystal X-ray structural data are available. A wide variety of complexes has been prepared and investigated by IR spectroscopy; hypsochromic shifts ranging from 20 to 110 cm^{-1} have been reported[81, 85, 91, 92]. In the particular case of transition metal halides such as VX_4, TiX_4, NbX_5 (X = Cl, Br), the mere dissolution of the salt in excess acetonitrile affords octahedral complexes of stoichiometry $M(CH_3CN)_nX_m$ (n + m = 6)[91].

Several transition metal carbonyls also form complexes with alkyl nitriles, where one, two, or three carbonyl groups are replaced by the nitrile[84]. The halide as well as the carbonyl complexes with acetonitrile ligands are excellent intermediates in the formation of new transition metal complexes not available by other routes[81, 91]. From spectroscopic data (d → d transitions), it has been concluded that, with regard to "ligand strength", the nitrile group is a better ligand than water or ammonia, occupying a high position in the spectrochemical series[66, 81].

Table 8. Stretching frequency ν_{CN} and bond length l_{CN} of some nitriles and of complexes thereof.

Compound	ν_{CN} (cm^{-1})	Ref.	l_{CN} (pm)	Ref.
CH_3CN	2267	Table 2	116	84)
$CH_3CN \cdot BF_3$	2355	85)	113	83)
$[ClAl(CH_3CN)_5][AlCl_4]_2$	2330	85)	114.9^a	84)
			110.5^b	84)
$NC(CH_2)_2CN$	2257	86)	116^c	87)
$[Cu \cdot NC(CH_2)_2CN]NO_3$	2283	86)	114	87)
$NC(CH_2)_3CN$	2249	86)	116^c	87)
$[Cu \cdot NC(CH_2)_3CN]NO_3$	2278	86)	114	88)
$NC(CH_2)_4CN$	2247	86)	116^c	87)
$[Cu \cdot NC(CH_2)_4CN]NO_3$	2273	86)	113	89)

^{a} Equatorial CH_3CN. ^{b} CH_3CN trans to Cl. ^{c} estimated, from other nitriles.

Table 8 includes copper(I) complexes of the dinitriles succinonitrile, glutaronitrile and adiponitrile. The crystal structure shows linear $C - C \equiv N \ldots Cu$ arrangements in the three cases; each copper ion is surrounded tetrahedrally by four nitrile groups. Evidently, the nitrile lone pair orbitals act as the donor part in these EDA complexes. Since Cu (I) has the electronic configuration $3d^{10}$, the 4s and 4p orbitals have to be considered as the acceptor orbitals (sp^3, tetrahedral).

In the case of succinonitrile, the crystal structure consists of polymeric chains of the following type[88]:

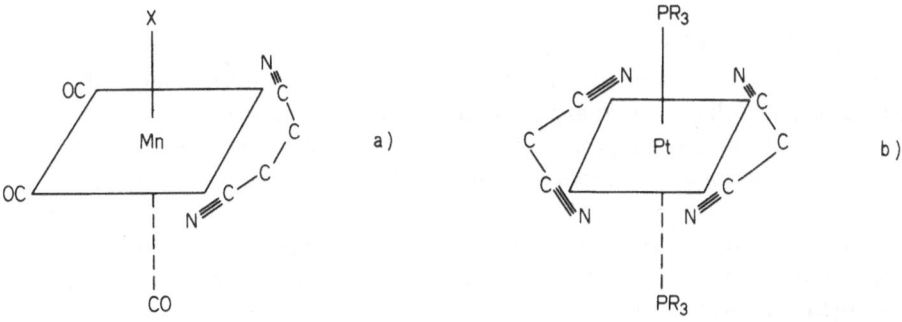

The crystals involving glutaronitrile consist of an infinite two-dimensional network[88], whereas adiponitrile gives rise to an infinite three-dimensional network[89]. In all cases the anion $(NO_3^- -)$ is embedded in the network.

Many other dinitrile complexes of Pt, Ni, Co, and Zn have been studied[93], and with a few exceptions to be discussed below, a bridging structure involving the lone pair orbitals of both nitrogen atoms was assumed, based on a hypsochromic shift, ν_{CN}, and on the absence of the stretching frequency of the free nitrile.

A few cases have been reported, where complex formation of a nitrile with a transition metal gave rise to a bathochromic shift of ν_{CN}. The first such case was $(Ni(0)CO(dialkylcyanamide)_2$, where the coordinated organic ligand shows a CN stretching frequency 200 cm^{-1} lower than that of the free ligand[94]. In analogy to olefins and acetylenes, where coordination with the π-orbitals leads generally to lowering of the C=C and C≡C stretching frequencies respectively[66], it was assumed that the bonding of the nitrile group is "acetylene-like" and involves the π-orbitals of the C≡N bond.

Based on the same argument, the structures shown in Fig. 9 have been suggested for a succinonitrile derivative of halogenpentacarbonyl manganese(I) ($\Delta\nu_{CN} =$ −185 cm^{-1})[95], and for the malononitrile derivative of a platinum(0) complex

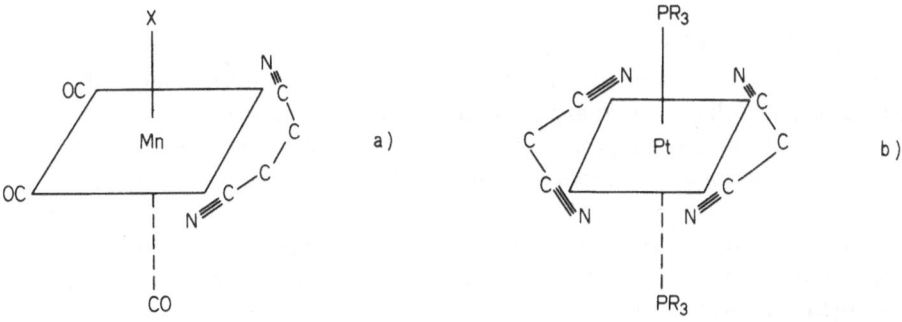

Fig. 9. Proposed structures for a) $Mn(CO)_3(NCCH_2CH_2CN)X$[95], and b) $Pt(P(C_6H_5)_3)_2(NCCH_2CN)_2$[93]

$(\Delta\nu_{CN} = -51 \text{ cm}^{-1})$[93]. For the manganese complex, the mononuclear structure was corroborated by molecular weight determination. Presumably, the geometric requirements for bidentate coordination through the two CN π-systems of the same dinitrile molecule are just matched in the case of succinonitrile and the Mn(I) ion, whereas with Cu(I) the lone pair interaction in combination with the stiff C−C≡N arrangement forces the same dinitrile ligand to form bridges between two metal centers[88]. Malononitrile appears to be specially suited for bidentate coordination via the two π-systems, because apart from the complex represented in Fig. 9b, similar complexes with Co, Ni, Pd, and also with Zn, have been reported[93]. Although the bathochromic shift of ν_{CN} appears to be a good evidence for the assumed structures, there is an obvious need for unambiguous single crystal X-ray structural data for these complexes.

Acrylonitrile appears to coordinate to transition metal ions primarily through its carbon-carbon double bond (highest occupied molecular orbital of that molecule[96]). This mode of coordination has been deduced from the absence of the stretching frequency $\nu_{C=C}$[97] and, in the cases of $Fe(CO)_4(CH_2=C-CN)$[98] and $Ni(PR_3)_2(CH_2=CH-CN)$[99], confirmed by X-ray structural analysis. The CN bond of the acrylonitrile ligand is slightly weakened in these complexes, as shown in a small increase of the bond length and decrease of ν_{CN}.

4.2 Complexes Involving Polyacrylonitrile

Polyacrylonitrile can be dissolved in concentrated aqueous solutions of $ZnCl_2$ (40−60%), and fibers can be spun from such solutions[100]. The interaction of nitrile lone pair electrons with Zn^{2+}, associated with the usual hypsochromic shift of ν_{CN} of 40−60 cm^{-1}, has been reported[93]. Evidently, this interaction suffices to lift the dipole-dipole interaction of the CN groups of the polymer, thus permitting dissolution. Stronger Lewis acids, such as VCl_4, $AlCl_3$, etc., have not found similar application, because they are not stable in aqueous solution.

The complex formation of Cu (I) ions with nitrile groups has found an interesting application for dyeing acrylonitrile fibers with anionic dyestuffs. The copper ion is readily absorbed by polyacrylonitrile and copolymers thereof, which then acquire almost unlimited affinity for many anionic dyes, permitting dyeing to very deep shades, including black[101, 103].

Since Cu (I) salts are either unstable or insoluble in water, the required valency state is obtained by in situ reduction of a Cu (II) salt, e. g., $CuSO_4$, preferably with hydroxylamine sulfate or hydrazine sulfate, in the dye bath. (Cu (II) itself is ineffective.) The reducing agent appears to form a weak complex with the Cu (I) ion, maintaining it in solution, but weak enough for the Cu (I) to be available for absorption by polyacrylonitrile. The metal ions are coordinatively bonded to the nitrile groups of the polymer. The original inorganic anion is electrostatically bonded to the copper-nitrile complex, and then gradually displaced by the slower diffusing, but more tightly bonded dyestuff anion.

If a polyacrylonitrile fiber is pretreated with Cu (I) salt, in the absence of dyestuff, the subsequent dye uptake, at comparable copper content, is greatly reduced. It was assumed[103] that the small Cu(I) ions and accompanying inorganic anions can diffuse into the more ordered regions of the fiber, where the large dyestuff anion cannot follow.

If, on the other hand, Cu (I) and dyestuff are applied simultaneously, a loose coordination compound is formed from Cu (I) and just one nitrile group[103]. The anion exchange is assumed to take place in this phase, essentially in the surface region of the fiber. Copper ion and dyestuff anion then diffuse together towards the inner regions of the fiber. In a later reaction phase, a more stable configuration is achieved in which the Cu (I) ion is surrounded by more than one nitrile group, ideally by four such groups in a tetrahedral arrangement (cf. the preceding section). Presumably, this last phase is slow, since the steric requirements for the formation of such a tetrahedral complex are stringent, and it will take some time until the diffusing Cu (I) ion finds a place where the conditions for the formation of the most stable complex are given. It appears evident that these conditions can be found only in the less ordered (amorphous) regions of the fiber.

5 Conclusions

The strongly polar nitrile goups are responsible for most of the noncovalent interactions of polyacrylonitrile, and essentially all characteristic properties of the polymer and of fibers therefrom can be traced back to such interactions.

For didactic (and partly for historic) reasons we have divided the noncovalent interactions in dipole-dipole interactions, hydrogen bonding and electron-donor-acceptor complexes. It should, however, be borne in mind that these phenomena are intimately related. Morokuma's method[32] to decompose the overall interaction energy into electrostatic, polarization, charge transfer and exchange repulsion components, has been of the greatest utility in elucidating these interactions. Through the efforts of many workers, electrostatic forces are presently believed to be the predominant factor, not only for dipole-dipole interactions[7, 8], but also for hydrogen bonding[32, 38] and for weak to medium electron-donor-acceptor complexes[32, 104]. In particular, energy and directionality of the intermolecular forces are generally well described by purely electrostatic models. This does not mean that the quantum mechanical forces, especially charge transfer and exchange repulsion, are negligible in all these cases; however, they tend to cancel each other, as discussed briefly in Section 3.1. In the particular case of transition metal complexes, on the other hand, charge transfer is generally assumed to play the dominant part in noncovalent bonding (see e. g.,[66]).

6 References

1. Hinrichsen, G.: Melliand Textilber. *58*, 7 (1977)
2. Handbook of Chemistry and Physics, CRC Press, 57 th Edition 1976

3. Saum, A. M.: J. Polymer Sci. *42*, 57 (1960)
4. Buckingham, A. D., Raab, R. E.: J. Chem. Soc. *1961*, 5511
5. Zhukova, E. L.: Opt. Spektroscop. *4*, 750 (1958)
6. Freedman, T. B., Nixon, E. R.: Spectrochim. Acta 28 A, 1375 (1972)
7. Paoloni, L., Hauser, S.: Bull. Soc. Chim. Belg. *84*, 219 (1975)
8. Dagnino, M. R., La Manna, G., Paoloni, L.: Chem. Phys. Letters *39*, 552 (1976), and references therein
9. Rowlinson, J. S.: Trans. Faraday Soc. *45*, 974 (1949)
10. Lambert, J. D., Roberts, A. H., Rowlinson, J. S., Wilkinson, V. J.: Proc. Royal Soc., A, *196*, 113 (1949)
11. Thomas, B. H., Orville-Thomas, W. J.: J. Mol. Structure *3*, 191 (1969)
12. Krigbaum, W. R., Takita, N.: J. Polymer Sci. *43*, 467 (1960)
13. Bohn, C. R., Schaefgen, J. R., Statton, W. O.: J. Polym. Sci. *55*, 531 (1961)
14. Rosenbaum, S.: J. Appl. Polymer Sci. *9*, 2071 (1965)
15. See, e. g., Stuart, H. A. (ed.): Die Physik der Hochpolymeren, Vol. 1. Berlin, Heidelberg, New York: Springer 1952
16. Dart, S. L.: Textile Res. J. *30*, 372 (1960)
17. a) Knudsen, J. P., Fitzgerald, W.: The Influence of Gel-Network Mechanics on the Tensile Properties of Wet Spun Fibers, Presented at Gordon Conference on Textiles, 1965; b) Bell, J. P., Dumbleton, J. H.: Textile Res. J. *41*, 196 (1971)
18. a) Hinrichsen, G., Orth, H.: Kolloid Z. *247*, 844 (1971); b) Hinrichsen, G.: J. Appl. Polymer Sci. *17*, 3305 (1973); c) Colvin, B. G., Storr, P.: Europ. Polymer J. *10*, 337 (1974)
19. Huang, Y. S., Koenig, J. L.: Appl. Spectrosc. *25*, 620 (1971)
20. See, e. g., Grassie, N., Hay, J. N.: J. Polymer Sci. *56*, 189 (1962); Friedlander, H. N., Peebles, L. H., Brandrup, J., Kirby, J. R.: Macromolecules *1*, 79 (1968); Ulbrich, J., Makschin, W.: Faserforsch. Textiltech. *22*, 381 (1971)
21. Dunn, P., Ennis, B. C.: J. Appl. Polymer Sci. *14*, 1795 (1970)
22. Hinrichsen, G.: Angew. Makromol. Chem. *20*, 121 (1974)
23. Schatzki, T. F.: J. Appl. Polymer Sci. *5*, S1 (1961)
24. Loshaek, S., Fox, T. G.: Bull. Am. Phys. Soc. *1*, 123 (1956)
25. Kimmel, R. M., Andrews, R. D.: J. Appl. Phys. *36*, 3063 (1965)
26. Ogura, K., Kawamura, S., Sobue, H.: Macromolecules *4*, 79 (1971)
27. Hinrichsen, G.: J. Polymer Sci. C, *38*, 303 (1972)
28. Cotton, G. R., Schneider, W. C.: Kolloid Z. *192*, 16 (1963)
29. Andrews, R. D., Kimmel, R. M.: Polymer Letters *3*, 167 (1965)
30. See, e. g., Murayama, T., Bell, J. P.: J. Polymer Sci. A–2 *8*, 437 (1970)
31. Rosenbaum, S.: J. Appl. Polymer, Sci. *9*, 2085 (1965)
32. Umeyama, H., Morokuma, K.: J. Amer. Chem. Soc. *99*, 1316 (1977) and references therein; Morokuma, K.: J. Chem. Phys. *55*, 1236 (1971)
33. Kollman, P. A., Allen, L. C.: J. Amer. Chem. Soc. *93*, 4997 (1971)
34. Al-Jabury, M., Turner, D. W.: J. Chem. Soc. *1964*, 4434
35. Field, F. H., Franklin, J. R.: Electron impact phenomena, New York: Academic Press 1957
36. Allred, A., Rochow, E.: J. Inorg. Nucl. Chem. *5*, 264, 269 (1958)
37. Pimentel, G., McClellan, A.: The hydrogen bond. Freeman, W. H., San Francisco, Calif. 1960
38. Kollman, P.: J. Amer. Chem. Soc. a) *99*, 4875 (1977); b) *100*, 2974 (1978)
39. Luck, W. A. P.: Topics Curr. Chem. *64*, 113 (1976)
40. Owicki, J. C., Scheraga, H. A.: J. Amer. Chem. Soc. *99*, 7403 (1977)
41. Swaminathan, S., Beveridge, D. L.: J. Amer. Chem. Soc. *99*, 8392 (1977)
42. Rahman, A., Stillinger, F. H.: J. Chem. Phys. *55*, 336 (1971); J. Amer. Chem. Soc. *95*, 7943 (1973)
43. Stillinger, F. H.: Adv. Chem. Phys. *31*, 1 (1975)
44. Hankins, D., Moskowitz, J., Stillinger, F.: J. Chem. Phys. *53*, 4544 (1970)
45. Del Bene, J. E.: J. Chem. Phys. *57*, 1899 (1972)
46. Popkie, H., Kistenmacher, H., Clementi, E.: J. Chem. Phys. *59*, 1325 (1973)

47. Stillinger, F., Rahman, A.: J. Chem. Phys. *60*, 1545 (1974)
48. Diercksen, G. H. F., Kraemer, W. P., Roos, B. O.: Theoret. Chim. Acta *36*, 249 (1975)
49. Kollman, P., McKelvey, J., Johansson, A., Rothenberg, S.: J. Amer. Chem. Soc. *97*, 955 (1975)
50. Matsuoka, O., Clementi, E., Yoshimine, M.: J. Chem. Phys. *64*, 1351 (1976)
51. Swietoslowsky, W.: Acad. Polon. *1931*, Nr. 2,6.7
52. Bernal, J. D., Fowler, R. H.: J. Chem. Phys. *1*, 515 (1933)
53. Dyke, T. R., Muenter, J. S.: J. Chem. Phys. *60*, 2929 (1974)
54. a) Perelygin, I. S., Shaikhova, A. B.: Opt. Spectrosk. *31*, 205 (1971)
    b) LeNarvor, A., Gentric, E., Saumagne, P.: Can. J. Chem. *49*, 1933 (1971)
55. Kobayashi, H.: Kagaku (Kyoto) *23*, 315 (1953)
56. Del Bene, J. E.: Chem. Phys. Letters. *24*, 203 (1974)
57. Murthy, A. S. N., Saini, G. R., Devi, K., Shah, S. B.: Adv. Mol. Relax. Proc. *7*, 255 (1975)
58. Turner, D. W., Baker, A. D., Baker, C., Brundle, C. R.: Molecular photoelectron spectroscopy, p. 346. New York: Wiley 1970
59. Siderova, A. I., Narziev, B. N.: Ukr. Fiz. Zh. *12*, 317 (1967)
60. Brandrup, J., Immergut, E. H. (eds.): Polymer Handbook, New York: Wiley 1975
61. See, e. g. Schulz, G. V.: Adv. Polymer Sci. *9*,1 (1972)
62. Dainton, F. S., Sisley, W. D.: Trans. Faraday Soc. *59*, 1369 (1963)
63. White, S. C., Thompson, H. W.: Proc. Roy. Soc. London, *A−291*, 460 (1966)
64. Henrici-Olivé, G., Olivé, S.: unpublished
65. Henrici-Olivé, G., Olivé, S.: Polymer Bulletin, *1*, 47 (1978)
66. See, e.g., Henrici-Olivé, G., Olivé, S.: Coordination and catalysis, Weinheim and New York: Verlag Chemie 1977
67. Bamford, C. H., Jenkins, A. D., Johnston, R.: Proc. Roy. Soc., London, *A−241*, 364 (1957)
68. Costain, C. C., Stoicheff, B. P.: J. Chem. Phys. *30*, 777 (1959)
69. Ger. Offen. 2,343,571, *1972* (DuPont)
70. Bell, J. P., Murayama, T.: J. Appl. Polymer Sci. *12*, 1795 (1968)
71. Rosenbaum, S.: Textile Res. J. *34*, 291 (1964)
72. Sotton, M., Vialard, A. M., Rabourdin, Ch.: Bull. Scient. ITF (France) *2*, 173 (1973)
73. Sotton, M., Jaquemart, J., Monrocq, R.: Bull. Scient. ITF (France), *2*, 247 (1973)
74. Stewart, J. C., Whewell, C. S.: Text. Res. J. *30*, 903, 912 (1960)
75. Friedlander, H. N., Menikheim, V.: Chemical reactions on polymeric fiber surfaces, in Ledwith, A., North, A. M. (eds.): Molecular behavior and the development of polymeric materials. London: Chapman and Hall 1975
76. Miller, B.: The wetting of fibers, in Schick, M. J. (ed.): Surface characteristics of fibers and textiles, Part. II. New York: Marcel Dekker 1977
77. Pontrelli, G. J.: Partial analysis of comfort's gestalt, in Hollies, N. R. S., Goldman, R. F. (eds.): Clothing comfort. Ann Arbor Sci. Publ. Inc. 1977
78. B. P. 851,829; 852,521, *1977*, (Bayer A. G.); see also Melliand Textilber. *58*, 11 (1977)
79. Gur-Arieh, Z., Ingamells, W., Peters, R. H.: J. Appl. Polymer Sci. *20*, 41 (1976); Gur-Arieh, Z., Ingamells, W.: J. S. D. C. *90*, 8 (1974)
80. U. S. P. 3,147,322 (1964)
81. Walton, R. A.: Quart. Rev. (London) *19*, 126 (1965)
82. Ha, T. K.: J. Mol. Structure *11*, 185 (1972)
83. Hoard, J. L., Owen, T. B., Buzzell, A., Salmon, O. N.: Acta, Cryst., *3*, 130 (1950)
84. Howard, J. A. K., Smart, L. E., Gilmore, C. J.: J. C. S. Chem. Comm. *1976*, 477
85. Purcell, K. F., Drago, R. S.: J. Amer. Chem. Soc. *88*, 919 (1966)
86. Matsubara, I.: Bull. Chem. Soc., Japan *34*, 1710, 1719 (1961): *35*, 27 (1962)
87. See, e. g., Cotton, F. A., Wilkinson, G.: Advanced inorganic chemistry, 3rd ed. New York: Interscience Publ. 1972
88. Kinoshita, Y. K., Matsubara, I., Saito, U.: Bull. Chem. Soc., Japan *32*, 741, 1216 (1959)
89. Kinoshita, Y. K., Matsubara, I., Higuchi, T., Saito, Y.: Bull. Chem. Soc., Japan *32*, 1221 (1959)
90. Brown, T. L., Kubota, H.: J. Amer. Chem. Soc. *83*, 4175 (1961)

91. Walton, R. A.: Progress Inorg. Chem. *16*, 1 (1972)
92. Reedijk, J., Zuur, A. P., Groeneveld, W. L.: Rec. Trav. Chim. Pays Bas. *86*, 1127 (1967)
93. Al-Janabi, M.: Thesis, University of Delaware 1968
94. Bock, H., tom Dieck, H.: Ber. *99*, 213 (1966)
95. Farona, M. F., Bremer, N. J.: J. Amer. Chem. Soc. *88*, 3735 (1966)
96. Mullen, P. A., Orloff, M. K., Theoret. Chim. Acta *23*, 278 (1971)
97. Schrauzer, G. N.: J. Amer. Chem. Soc. *81*, 5310 (1959)
98. Luxmore, A. R., Truter, M. R.: Acta Cryst. *15*, 1117 (1962)
99. Guggenberger, L. J.: Inorg. Chem. *12*, 499 (1973)
100. U. S. 3,346,685, *1967* (Dow Chemical Company)
101. Feild, T. A., Fremon, G. H.: Textile Research J. *21*, 531 (1951)
102. Blaker, R. H., Katz, S. M., Laucius, J. F., Remington, W. R., Schroeder, H. E.: Discuss. Faraday Soc. *16*, 210 (1954)
103. Rath, H., Rehm, H., Rummler, H., Specht, E.: Melliand Textilber. *38*, 431 (1957)
104. Lathan, W. A., Morokuma, K.: J. Amer. Chem. Soc. *97*, 3615 (1975)

Received January 8, 1979

Author Index Volumes 1–32

Polymers

Properties and Applications

Editorial Board:
H.-J. Cantow, H. J. Harwood, J. P. Kennedy, A. Ledwith, J. Meißner, S. Okamura, G. Olivé, S. Olivé

Volume 1
B. Rånby, J. F. Rabek

ESR Spectroscopy in Polymer Research

1977. 356 figures, 29 tables. XIV, 410 pages
ISBN 3-540-08151-8

Volume 2
H.-H. Kausch

Polymer Fracture

1978. 180 figures, 23 tables. X, 332 pages
ISBN 3-540-08786-9

Volume 3
A. Knop, W. Scheib

Chemistry and Application of Phenolic Resins

1979. 111 figures, 88 tables. XIII, 269 pages
ISBN 3-540-09051-7

Springer-Verlag
Berlin
Heidelberg
New York

Polymer Bulletin

Editors:

Prof. H.-J. Cantow, Institute for Macromolecular Chemistry, University of Freiburg, Stefan-Meier-Straße 31, D-78 Freiburg/Germany

Prof. J. P. Kennedy, Dept. of Polymer Science, The University of Akron, Akron, OH 44325/USA

Prof. T. Saegusa, Dept. of I ynthetic Chemistry, Kyoto University, Kyoto, 606 Japan

The articles are to be sent to one of the editors or to Springer-Verlag Berlin Heidelberg New York

Polymer Bulletin

Preface

To cope with the rapid progress of polymer science, a new journal is now published characterized by emphasis on rapid publication of papers containing a most concise description of results.
The character of the new journal is between the purely archival journal of full papaers and the so-called "letter journals" consisting exclusively of short communications.

Ask for our detailed leaflet!

The journal consists of one volume a year, published in 12 issues.

Subscription information upon request.